Frontiers in Mathematics

Ilaria Cardinali
Stanley E. Payne

q-Clan Geometries in Characteristic 2

Birkhäuser Verlag
Basel · Boston · Berlin

Authors:

Ilaria Cardinali
Department of Mathematics
University of Siena
Pian dei Mantellini, 44
53100 Siena
Italy
e-mail: cardinali3@unisi.it

Stanley E. Payne
Department of Mathematical Sciences
University of Colorado at Denver
and Health Sciences Center
Campus Box 170
P.O. Box 173364
Denver, CO 80217-3364
USA
e-mail: stanpayne@mac.com

2000 Mathematical Subject Classification: 51E12, 51E21

Library of Congress Control Number: 2007929014

Bibliographic information published by Die Deutsche Bibliothek
Die Deutsche Bibliothek lists this publication in the Deutsche Nationalbibliografie; detailed bibliographic data is available in the Internet at <http://dnb.ddb.de>.

ISBN 978-3-7643-8507-1 Birkhäuser Verlag, Basel - Boston - Berlin

Part of Springer Science+Business Media
Cover design: Birgit Blohmann, Zürich, Switzerland
Printed on acid-free paper produced from chlorine-free pulp. TCF ∞

ISBN 978-3-7643-8507-1

ISBN 978-3-7643-8508-8 (eBook)

9 8 7 6 5 4 3 2 1

www.birkhauser.ch

Contents

Preliminaries

Introduction

This memoir is a thoroughly revised and updated version of the *Subiaco Notebook*, which since early 1998 has been available on the web page of the senior author at `http://www-math.cudenver.edu/~spayne/`.

Our goal is to give a fairly complete and nearly self-contained treatment of the known (infinite families of) generalized quadrangles arising from q-clans, i.e., flocks of quadratic cones in $PG(3, q)$, with $q = 2^e$. Our main interest is in the construction of the generalized quadrangles, a determination of the associated ovals, and then a complete determination of the groups of automorphisms of these objects. A great deal of general theoretical material of related interest has been omitted. However, we hope that the reader will find the treatment here to be coherent and complete as a treatment of one major part of the theory of flock generalized quadrangles.

Since the appearance of the Subiaco Notebook the Adelaide q-clans have been discovered, generalizing the few examples of "cyclic" q-clans found first by computer (see [PPR97]). The revised treatment given here of the cyclic q-clans, which is a slight improvement of that given in [COP03], allows much of the onerous computation in the Subiaco Notebook to be avoided while at the same time allowing a more unified approach to the general subject. However, a great deal of computation is still unavoidable.

Most of the work done on the Adelaide examples, especially our study of the Adelaide ovals, and major steps in the clarification of the connection between the so-called Magic Action of O'Keefe and Penttila (see [OP02]) and the Fundamental Theorem of q-clan geometry, took place while the senior author was a visiting research professor at the Universities of Naples, Italy, and Ghent, Belgium, during the winter and spring of the year 2002. This was made possible by a semester-long sabbatical provided by the author's home institution, the University of Colorado at Denver.

During the two months he spent in Italy, at the invitation of Professor Dr. Guglielmo Lunardon (with the collaboration of Prof. Laura Bader), he received generous financial support from the GNSAGA and the University of Naples, along with a great deal of personal support from his colleagues there. Also, much of the material on the cyclic q-clans derives from the reports [Pa02a] and [Pa02b] and has appeared in [CP03].

During his two months in Belgium, at the invitation of Prof. Dr. Joseph A. Thas, he was generously supported by the Research Group in Incidence Geometry at Ghent University. As always, it was a truly great pleasure to work in the stimulating and friendly atmosphere provided by his colleagues there. All the material on the Adelaide ovals was adapted from [PT05].

It was in Naples during the trimester February-March 2002 that the second author became familiar with the Four Lectures in Naples [Pa02a] and the idea of working with the senior author to complete the present memoir first occured to us.

The final steps in the clarification of the connections between the Magic action and the Fundamental Theorem, together with a revision of all the extracts from the Subiaco Notebook, were taken while the second author was visiting the Department of Mathematics at the University of Colorado at Denver in the Fall of 2002. During that period the present work was essentially completed. It was a great pleasure to study and work in the friendly, enjoyable and stimulating atmosphere provided by all our colleagues at the University of Colorado at Denver. We wish to thank in particular our friend and colleague Dr. William Cherowitzo for several fruitful mathematical conversations and for his many suggestions concerning the use of LaTeX. We hope that this memoir will become a standard source of information for those who are interested in q-clan geometries when q is even.

Finite Generalized Quadrangles

For a thorough introduction to finite generalized quadrangles, see the monograph [PT84]. However, for the convenience of the reader we review here a few definitions and elementary results without proof.

A *finite generalized quadrangle* (GQ) is an incidence structure $\mathcal{S} = (\mathcal{P}, \mathcal{B}, I)$ in which $\mathcal{P}$ and $\mathcal{B}$ are disjoint (finite, non-empty) sets of objects called *points* and *lines*, respectively, and for which $I \subseteq (\mathcal{P} \times \mathcal{B}) \cup (\mathcal{B} \times \mathcal{P})$ is a symmetric point-line incidence relation satisfying the following axioms:

(i) Each point is incident with $t+1$ lines ($t \geq 1$) and two distinct points are incident with at most one line;

(ii) Each line is incident with $s+1$ points ($s \geq 1$) and two distinct lines are incident with at most one point;

(iii) If x is a point and L is a line not incident with x, there is a unique pair $(y, M) \in \mathcal{P} \times \mathcal{B}$ for which $xIMIyIL$.

The integers s and t are the *parameters* of the GQ, and $\mathcal{S}$ is said to have *order* (s,t). If $s = t$, $\mathcal{S}$ is said to have order s. If $\mathcal{S}$ has order (s,t), then $|\mathcal{P}| = (1+s)(1+st)$

and $|\mathcal{B}| = (1+t)(1+st)$. If $\mathcal{S}$ is a GQ of order (s,t), then the incidence structure $\mathcal{S}^D$ having as *lines*, the points of $\mathcal{S}$ and as *points* the lines of $\mathcal{S}$, is a GQ of order (t,s) called the *point-line dual* of $\mathcal{S}$.

The classical GQs of order q are $W(q)$, which arise as the absolute points and lines of a symplectic polarity of $PG(3,q)$; and $Q(4,q)$, which arise as the points and lines of a non-singular quadric of $PG(4,q)$. By ([PT84], 3.2.1), $W(q)$ is isomorphic to the dual of $Q(4,q)$ and is self-dual for q even. Another class of GQs of order q are the $T_2(\mathcal{O})$ constructed by J. Tits and first appearing in [De68] (see [PT84], 3.1.2). Let $\mathcal{O}$ be an oval of $PG(2,q)$, and let $\pi_\infty = PG(2,q)$ be embedded as a hyperplane in $PG(3,q)$. The *points* of $T_2(\mathcal{O})$ are: (i) the points of $PG(3,q) \setminus \pi_\infty$, called the *affine points*; (ii) the planes of $PG(3,q)$ meeting π_∞ in a single point of $\mathcal{O}$; and (iii) a symbol (∞). The *lines* of $T_2(\mathcal{O})$ are: (a) the lines of $PG(3,q)$, not in π_∞, that meet π_∞ in a single point of $\mathcal{O}$; and (b) the points of $\mathcal{O}$. Incidence is as follows: a point of type (i) is incident only with the lines of type (a) that contain it; a point of type (ii) is incident with all the lines of type (a) contained in it and with the unique line of type (b) on it; the point of type (iii) is incident with all lines of type (b) and no line of type (a).

The GQ $T_2(\mathcal{O})$ has a slightly simpler description that is easily seen to be equivalent to the one given above *when q is even*. Let N be the nucleus of $\mathcal{O}$. Here the points of $T_2(\mathcal{O})$ are the points of $PG(3,q) \setminus \pi_\infty$ and the planes of $PG(3,q)$ that contain the point N. The lines of $T_2(\mathcal{O})$ are the lines of $PG(3,q)$ that meet $\mathcal{O}$ in a single point. Incidence is that inherited from $PG(3,q)$. For example, the plane π_∞ as a point of $T_2(\mathcal{O})$ is incident with the lines of π_∞ that are tangent to $\mathcal{O}$, i.e., the lines of π_∞ incident with N.

There is a natural analog $T_3(\Omega)$ of the preceding construction also given by J. Tits that yields GQ of order (q,q^2) starting from an ovoid Ω in $PG(3,q)$ (i.e., q^2+1 points no three on a line). Let Ω be an ovoid of $\Sigma = PG(3,q)$ which is embedded as a hyperplane in $PG(4,q)$. Construct the point-line incidence geometry $\mathcal{S} = (\mathcal{P}, \mathcal{B}, I)$ with pointset $\mathcal{P}$, lineset $\mathcal{B}$, and incidence I as follows:

The points of $\mathcal{S}$, i.e., elements of $\mathcal{P}$, are of three types:

(i) The points of $PG(4,q) \setminus \Sigma$;

(ii) The hyperplanes of $PG(4,q)$ meeting Σ in a plane tangent to $\mathcal{O}$.

(iii) The symbol (∞).

The lines of $\mathcal{S}$ are of two types:

(a) The lines of $PG(4,q)$ that are not contained in Σ and meet Σ in a (necessarily unique) point of Ω.

(b) The points of Ω.

Incidence in $T_3(\Omega)$ is defined by the following: the point (∞) is incident with the $1+q^2$ lines of type (b). Suppose $\triangle$ is a solid of $PG(4,q)$ meeting Σ in the plane T_p tangent to Ω at the point p. Then $\triangle$ is incident with p (as a line of type (b)) and with the q^2 lines of $\triangle$ through p but not contained in Σ. Each point x of $PG(4,q) \setminus \Sigma$ is incident with the $1+q^2$ lines px, $p \in \Omega$.

These GQ constructed by J. Tits first appeared in [De68] (also see [PT84]).

The GQ that are of primary interest in this memoir are those with parameters (q^2, q) arising from a q-clan, i.e., from a flock of a quadratic cone in $PG(3,q)$. These will be introduced in Chapter 1. If Ω is an elliptic quadric in Σ, it turns out that $T_3(\Omega)$ is isomorphic to the point-line dual of a GQ arising from a *linear flock*.

As this monograph is an extended and updated version of the Subiaco Notebook, it seems appropriate to present here an updated version of the Prolegomena from the Subiaco Notebook.

Prolegomena

At a conference in Han-Sur-Lesse, Belgium, held in 1979, W. M. Kantor [Ka80] surprised the finite geometry community with the construction of a family of generalized quadrangles (GQ) of order (q^2, q) for each prime power q with $q \equiv 2 \pmod 3$. His discovery of these examples began with a classical generalized hexagon, but his description of them introduced the following very important general method.

Let G be a finite group of order s^2t, $1 < s$, $1 < t$, together with a family $\mathcal{J} = \{A_i : 0 \leq i \leq t\}$ of $1+t$ subgroups of G, each of order s. In addition, suppose that for each $A_i \in \mathcal{J}$ there is a subgroup A_i^* of G, of order st, containing A_i. Put $\mathcal{J}^* = \{A_i^* : 0 \leq i \leq t\}$ and define as follows a point-line geometry $\mathcal{S} = (P, B, I) = \mathcal{S}(G, \mathcal{J})$ with pointset P, lineset B, and incidence I.

Points are of three kinds:

(i) the elements of G;
(ii) the right cosets A_i^*g, (for all $A_i^* \in \mathcal{J}^*$, $g \in G$);
(iii) a symbol (∞).

Lines are of two kinds:

(a) the right cosets A_ig, (for all $A_i \in \mathcal{J}$, $g \in G$);
(b) the symbols $[A_i]$, $A_i \in \mathcal{J}$.

A point g of type (i) is incident with each line A_ig, $A_i \in \mathcal{J}$; a point A_i^*g of type (ii) is incident with $[A_i]$ and with each line A_ih contained in A_i^*g; the point (∞) is incident with each line $[A_i]$ of type (b).

Kantor [Ka80] was the first to recognize that $\mathcal{S}(G, \mathcal{J})$ is a GQ of order (s,t) if and only if the following two conditions are satisfied:

K1: $A_iA_j \cap A_k = \{e\}$, for distinct i, j, k, and

K2: $A_i^* \cap A_j = \{e\}$, for $i \neq j$.

If the conditions K1 and K2 are satisfied, then

$$A_i^* = \cup\{A_i g : A_i g = A_i \text{ or } A_i g \cap A_j = \emptyset \text{ for all } A_j \in \mathcal{J}\},$$

so that A_i^* is uniquely defined by A_i and $\mathcal{J}$.

Suppose K1 and K2 are satisfied. For any $h \in G$ define θ_h by $g^{\theta_h} = gh$, $(A_i g)^{\theta_h} = A_i gh$, $(A_i^* g)^{\theta_h} = A_i^* gh$, $[A_i]^{\theta_h} = [A_i]$, $(\infty)^{\theta_h} = (\infty)$, for $g \in G$, $A_i \in \mathcal{J}$, $A_i^* \in \mathcal{J}^*$. Then θ_h is an automorphism of $\mathcal{S}(G, \mathcal{J})$ which fixes the point (∞) and all lines of type (b). If $G' = \{\theta_h : h \in G\}$, then clearly G' is a group isomorphic to G, and G' acts regularly on the points of type (i).

If K1 and K2 are satisfied, we say that $\mathcal{J}$ is a *4-gonal family for* G, or that $(G, \mathcal{J}, \mathcal{J}^*)$ is a *Kantor family.*

The appearance of the new GQ in Kantor [Ka80] inspired the development in Payne [Pa80] of a more specific recipe for constructing GQ. Then in Kantor [Ka86] the conditions in Payne [Pa80] for q odd were shown to be equivalent to having what today we call a q-clan. This in turn inspired an analogous interpretation in Payne [Pa85] for $q = 2^e$, and the discovery of a new infinite family of GQ with order (q^2, q), $q = 2^e$, e odd. The term q-*clan* was first used in Payne [Pa89]. When $q = 2^e$, along with each q-clan comes a collection of ovals in $PG(2, q)$. The ovals in Payne [Pa85] were new, but only exist for q a nonsquare. It was still true that the only examples of ovals not related to conics that were known for q a square were associated with the translation ovals (first constructed by B. Segre [Se57]) or with the single example of Lunelli-Sce [LS58] in $PG(2, 16)$. Then in 1993 W. Cherowitzo, T. Penttila, I. Pinneri and G. Royle discovered the Subiaco q-clans with their new ovals and GQ. By the time their paper [CPPR96] appeared, a great deal of work on the Subiaco geometries had been done. In particular we mention [Pa94], [BLP94], [PPP95], and [OKT96]. Work on the Subiaco GQ also directly inspired [Pa96] and [Pa95], and the latter contains additional results on the Subiaco ovals.

With the computationally efficient tensor product notation of [Pa95] having been found only after most the computations for the Subiaco GQ had been done, with so much work on the Subiaco GQ involving computations only sketched in the published articles, and with the discovery of the Adelaide examples, we feel that there should exist a single, coherent treatment of the general body of work related especially to the GQ arising from q-clans and their associated ovals. Especially we wanted to see one treatment determining the collineation groups of the GQ and of the associated ovals. Hence these notes!

The scope of this replacement for the Subiaco Notebook remains similar to that of the Subiaco Notebook (but of course the Adelaide examples were not known when the Subiaco Notebook was written). We have included an essentially self-contained introduction to q-clan geometry (i.e., flock GQ, etc.) with $q = 2^e$, giving a detailed construction of the known examples and a determination of their groups. Moreover, even though we attempt to give many details here and always enough for the energetic reader to fill in what we have left out, still there are many computations that we have found it feasible only to sketch.

The other geometries associated with the Subiaco q-clans, for example spreads of $PG(3, q)$ and translation planes, are also mentioned briefly, but the main goal is to give a complete, self-contained treatment of the GQ and ovals.

For a broad survey of these topics, including geometries over finite and infinite fields of arbitrary characteristic, see the survey by N. L. Johnson and S. E. Payne [JP97]. For an excellent survey of the geometries related to finite q-clans, especially flocks of quadratic cones known prior to the discovery of the Subiaco examples, see Chapers 7 and 9 by J. A. Thas in the *Handbook* [Bu95].

The most exciting development not covered in the present work is no doubt the geometric construction by J. A. Thas of flock GQ for all characteristics, along with his work on Property (G). See [Th98] and its references for this major work that succeeded in proving that when q is odd, Property (G) at a point completely characterizes the flock GQ. His work left the characteristic 2 case not quite finished, but more recently M. R. Brown (see [BBP06], [Br07] and [Br06]) has finished this case as well, giving new insight to the constructions.

The present work is not really an introduction to the general theory of finite generalized quadrangles per se, or even to *elation generalized quadrangles.* And although in the preceding section we have repeated Kantor's construction of elation generalized quadrangles as group coset geometries, we assume that the reader has access to some other source of basic concepts and proofs related to finite GQ, for example the monograph by S. E. Payne and J. A. Thas [PT84]. The most thorough recent treatment of related topics is surely the new monograph [TTVM06].

Chapter 1

q-Clans and Their Geometries

1.1 Anisotropism

Let $q = 2^e$, $F = GF(q)$. Let $A = \left(\begin{smallmatrix} x & y \\ w & z \end{smallmatrix}\right)$ and $A' = \left(\begin{smallmatrix} x' & y' \\ w' & z' \end{smallmatrix}\right)$ be arbitrary 2×2 matrices over F.

Definition 1.1.1. *$A \equiv A'$ provided $x = x'$, $z = z'$ and $y + w = y' + w'$.*

The definition of "$\equiv$" is chosen so that $A \equiv A'$ iff $\alpha A \alpha^T = \alpha A' \alpha^T$ for all $\alpha \in F^2$. Clearly "$\equiv$" is an equivalence relation.

Definition 1.1.2. *The matrix A is said to be* anisotropic *provided $\alpha A \alpha^T = 0$ iff $\alpha = (0, 0)$.*

Observation 1.1.3. Let A, A' be as above; let M be any 2×2 matrix over F; let $B \in GL(2, q)$; and let $\sigma \in Aut(F)$. Then

(i) $A \equiv A'$ implies that A is anisotropic iff A' is anisotropic.

(ii) A is anisotropic iff $BA^\sigma B^T$ is anisotropic.

(iii) $A - A'$ is anisotropic iff $[BA^\sigma B^T + M] - [B(A')^\sigma B^T + M]$ is anisotropic.

Recall the *absolute trace function* $tr\colon GF(q) \mapsto GF(2)$, $tr(x) = \sum_{i=1}^{e} x^{2^{i-1}}$.

Observation 1.1.4. The following are equivalent:

(i) $A = \left(\begin{smallmatrix} x & y \\ w & z \end{smallmatrix}\right)$ is anisotropic.

(ii) $a^2x + ab(y + w) + b^2z$ is irreducible over F as a homogeneous polynomial of degree 2 in a and b.

(iii) $tr\left(\dfrac{xz}{(y + w)^2}\right) = 1$.

1.2 q-Clans

Let $X : t \mapsto x_t$, $Y : t \mapsto y_t$, $Z : t \mapsto z_t$ be three functions from F to F, and put $\mathcal{C} = \{A_t \equiv \left(\begin{smallmatrix} x_t & y_t \\ 0 & z_t \end{smallmatrix}\right) : t \in F\}$. So $\mathcal{C}$ is a set of at most q equivalence classes of 2×2 matrices over F.

Definition 1.2.1. *$\mathcal{C}$ is a q-clan provided all pairwise differences $A_s - A_t$ $(s, t \in F,\ s \neq t)$ are anisotropic.*

Observation 1.2.2. We may assume that q-clan matrices are upper triangular, i.e., $\left(\begin{smallmatrix} x & y \\ w & z \end{smallmatrix}\right) \equiv \left(\begin{smallmatrix} x & y+w \\ 0 & z \end{smallmatrix}\right)$, and if $B = \left(\begin{smallmatrix} a & b \\ c & d \end{smallmatrix}\right)$, then the following holds:

$$B \begin{pmatrix} x & y \\ 0 & z \end{pmatrix} B^T \equiv \begin{pmatrix} a^2x + aby + b^2z & (ad + bc)y \\ 0 & c^2x + dcy + d^2z \end{pmatrix}. \tag{1.1}$$

Note that if $\mathcal{C}$ is a q-clan, then it does consist of q distinct equivalence classes. Since most of the time we write matrices of $\mathcal{C}$ in upper triangular form (since $q = 2^e$), usually we just write $A_t = \left(\begin{smallmatrix} x_t & y_t \\ 0 & z_t \end{smallmatrix}\right)$.

By Observation 1.1.4 the anisotropic condition is exactly that for any distinct $s, t \in GF(q)$, $tr((x_t + x_s)(z_t + z_s)/(y_t + y_s)^2) = 1$. It is easy to check that when $\mathcal{C}$ is a q-clan, each of the functions X, Y, Z is a permutation. Moreover, by Observation 1.1.3, if $0 \neq \mu \in F$, $\sigma \in Aut(F)$, $B \in GL(2, q)$, and if $x, y, z \in F$, replacing $A_t \in \mathcal{C}$ with $A'_t \equiv \mu B A_t^\sigma B^T + \left(\begin{smallmatrix} x & y \\ 0 & z \end{smallmatrix}\right)$ gives a new q-clan $\mathcal{C}' = \{A'_t : t \in F\}$.

Definition 1.2.3. *Let $\mathcal{C} = \{A_t : t \in F\}$ and $\mathcal{C}' = \{A'_t : t \in F\}$ be q-clans. $\mathcal{C}$ and $\mathcal{C}'$ are equivalent and we write $\mathcal{C} \sim \mathcal{C}'$ provided there exist the following: $0 \neq \mu \in F$, $B \in GL(2, q)$, $\sigma \in Aut(F)$, M a 2×2 matrix over F, and a permutation $\pi : t \mapsto \bar{t}$ on F for which the following holds:*

$$A'_{\bar{t}} \equiv \mu B A_t^\sigma B^T + M \text{ for all } t \in F. \tag{1.2}$$

1.3 Flocks of a Quadratic Cone

Let $K = \{(x_0, x_1, x_2, x_3) \in PG(3, q) : x_1^2 = x_0x_2\}$. So K is a quadratic cone in $PG(3, q)$ with vertex $V = (0, 0, 0, 1)$. Keep in mind that all quadratic cones of $PG(3, q)$ are equivalent under the action of $P\Gamma L(4, q)$ hence we can take our favorite one without loss of generality.

Definition 1.3.1. *A flock of K is a partition $\mathcal{F} = \{C_t : t \in F\}$ of $K \setminus \{V\}$ into q pairwise disjoint conics C_t.*

Each conic $C_t \in \mathcal{F}$ is a plane intersection $C_t = \pi_t \cap K$, where $\pi_t = [x_t, y_t, z_t, 1]^T$ is a plane not containing the vertex V. The notation $[x_t, y_t, z_t, 1]^T$ stands for $x_tX_0 + y_tX_1 + z_tX_2 + X_3 = 0$. We also refer to π_t as a *plane of the flock* $\mathcal{F}$. In fact, it has also become common just to define a flock to be the set of planes whose intersections with the cone are the conics that partition the cone. J. A. Thas [Th87] first proved the following:

Theorem 1.3.2 (J. A. Thas [Th87]). *Let $X : t \mapsto x_t$, $Y : t \mapsto y_t$, $Z : t \mapsto z_t$ be three functions on F. For $t \in F$, put $\pi_t = [x_t, y_t, z_t, 1]^T$, $C_t = \pi_t \cap K$, $\mathcal{F} = \{C_t : t \in F\}$. Also put $A_t \equiv \left(\begin{smallmatrix} x_t & y_t \\ 0 & z_t \end{smallmatrix}\right)$ and $\mathcal{C} = \{A_t : t \in F\}$. Then $\mathcal{C}$ is a q-clan iff $\mathcal{F}$ is a flock.*

Proof. $\mathcal{F}$ is a flock by definition if and only if any two distinct conics C_s and C_t with s different from t are disjoint. If $\pi_t = [x_t, y_t, z_t, 1]^T$ and $\pi_s = [x_s, y_s, z_s, 1]^T$ are the planes defining the two conics C_s and C_t, we want that π_s and π_t meet in a line which is external to K. Thus the system

$$\begin{cases} x_t X_0 + y_t X_1 + z_t X_2 + X_3 = 0 \\ x_s X_0 + y_s X_1 + z_s X_2 + X_3 = 0 \\ x_1^2 = x_0 x_2 \end{cases}$$

must have only the trivial solution. By subtracting the second equation from the first and making the appropriate substitutions, it is easy to see that the previous system has a non-trivial solution if and only if the following equation has a non-trivial solution in Y:

$$(x_t - x_s)Y^2 + (y_t - y_s)Y + (z_t - z_s) = 0.$$

Since we are in even characteristic, this holds if and only if $\frac{(x_t - x_s)(z_t - z_s)}{(y_t - y_s)^2}$ has absolute trace equal to 1 for any $s, t \in F$, $s \neq t$.

By Observation 1.1.4 this condition is equivalent to requiring that the matrix $A_s - A_t$ be anisotropic for any $s, t \in F$, $s \neq t$, which is exactly the condition for $\mathcal{C}$ to be a q-clan. □

Definition 1.3.3. *Two* flocks *are* projectively equivalent *when there exists a projective semilinear map of $PG(3, q)$ leaving the cone invariant and mapping one flock to the other.*

From now on let $\mathcal{C} = \{A_t \equiv \left(\begin{smallmatrix} x_t & y_t \\ 0 & z_t \end{smallmatrix}\right) : t \in F\}$ be a q-clan. Let $\mathcal{F}(\mathcal{C}) = \{\pi_t \cap K : \pi_t = [x_t, y_t, z_t, 1]^T : t \in F\}$ be the corresponding flock. Then $\mathcal{C}' = \{A'_t \equiv A_t - A_0 : t \in F\}$ is a q-clan equivalent to $\mathcal{C}$ with $A'_0 = \left(\begin{smallmatrix} 0 & 0 \\ 0 & 0 \end{smallmatrix}\right)$. And $T : [x, y, z, 1]^T \mapsto [x - x_0, y - y_0, z - z_0, 1]^T$ determines a projective (linear) map on $PG(3, q)$ (defined on planes not through V) that leaves invariant the cone K and maps $\mathcal{F}(\mathcal{C})$ to $\mathcal{F}(\mathcal{C}')$. The plane π'_0 of the flock $\mathcal{F}(\mathcal{C}')$ is the plane $\pi'_0 = [0, 0, 0, 1]^T$. Hence from now on we assume without loss of generality that each q-clan $\mathcal{C}$ contains the zero matrix, hence the corresponding flock $\mathcal{F}(\mathcal{C})$ contains a conic in the plane $\pi = [0, 0, 0, 1]^T$. It is possible that on certain occasions the matrices of $\mathcal{C}$ are labeled so that $A_s \in \mathcal{C}$ is the zero matrix for some nonzero $s \in F$, but usually this is not the case, i.e., usually $A_0 = \left(\begin{smallmatrix} 0 & 0 \\ 0 & 0 \end{smallmatrix}\right)$.

Theorem 1.3.4 (Cor. 7.14, p. 158 of J.W.P. Hirschfeld [Hi98]). *The most general projective semilinear map T of $PG(3, q)$ defined as a map on the planes of $PG(3, q)$ not through the vertex V, that leaves invariant the cone K, is given by the following. There is a matrix $B = \left(\begin{smallmatrix} a & b \\ c & d \end{smallmatrix}\right) \in GL(2, q)$, $0 \neq \lambda \in F$, $\sigma \in Aut(F)$, and there* *are fixed elements $x_0, y_0, z_0 \in F$, for which*

$$T: \begin{bmatrix} x \\ y \\ z \\ 1 \end{bmatrix} \longmapsto \begin{bmatrix} x' \\ y' \\ z' \\ 1 \end{bmatrix} = \begin{pmatrix} \lambda a^2 & \lambda ab & \lambda b^2 & x_0 \\ 0 & \lambda(ad+bc) & 0 & y_0 \\ \lambda c^2 & \lambda cd & \lambda d^2 & z_0 \\ 0 & 0 & 0 & 1 \end{pmatrix} \begin{bmatrix} x^\sigma \\ y^\sigma \\ z^\sigma \\ 1 \end{bmatrix}. \tag{1.3}$$

Proof. See [Hi98]. □

Theorem 1.3.4 has as an immediate consequence the following theorem which is central to the theory.

Theorem 1.3.5. *Let $\mathcal{C} = \{A_t \equiv \begin{pmatrix} xt & yt \\ 0 & zt \end{pmatrix} : t \in F\}$, $\mathcal{C}' = \left\{A'_t \equiv \begin{pmatrix} x'_t & y'_t \\ 0 & z'_t \end{pmatrix} : t \in F\right\}$ be two (not necessarily distinct) q-clans, normalized so that $A_0 = A'_0$ is the zero matrix, with corresponding flocks $\mathcal{F}(\mathcal{C})$ and $\mathcal{F}(\mathcal{C}')$. Then $\mathcal{F}(\mathcal{C})$ and $\mathcal{F}(\mathcal{C}')$ are projectively equivalent iff there exist the following:*

(i) $0 \neq \lambda \in F$,

(ii) $B = \begin{pmatrix} a & b \\ c & d \end{pmatrix} \in GL(2, q)$,

(iii) $\sigma \in Aut(F)$,

(iv) $\pi : F \to F : t \mapsto \bar{t}$, *a permutation, such that the following condition is satisfied:*

$$\begin{bmatrix} x'_{\bar{t}} \\ y'_{\bar{t}} \\ z'_{\bar{t}} \\ 1 \end{bmatrix} = \begin{pmatrix} \lambda a^2 & \lambda ab & \lambda b^2 & x'_{\bar{0}} \\ 0 & \lambda(ad+bc) & 0 & y'_{\bar{0}} \\ \lambda c^2 & \lambda cd & \lambda d^2 & z'_{\bar{0}} \\ 0 & 0 & 0 & 1 \end{pmatrix} \begin{bmatrix} x^\sigma_t \\ y^\sigma_t \\ z^\sigma_t \\ 1 \end{bmatrix}, \quad t \in F. \tag{1.4}$$

As $q = 2^e$ and the matrices of $\mathcal{C}$ and $\mathcal{C}'$ are upper triangular, Eq. (1.4) is equivalent to

$$A'_{\bar{t}} \equiv \lambda B A^\sigma_t B^T + A'_{\bar{0}}, \ t \in F. \tag{1.5}$$

Clearly Theorem 1.3.5 is just a very explicit way of saying that $\mathcal{C} \sim \mathcal{C}'$ iff $\mathcal{F}(\mathcal{C})$ and $\mathcal{F}(\mathcal{C}')$ are projectively equivalent.

1.4 4-Gonal Families from q-Clans

Put $P = \begin{pmatrix} 0 & 1 \\ 1 & 0 \end{pmatrix}$, and for $\alpha, \beta \in F^2$ define $\alpha \circ \beta$ by

$$\alpha \circ \beta = \alpha P \beta^T. \tag{1.6}$$

Then $(\alpha, \beta) \mapsto \alpha \circ \beta$ is a nonsingular, alternating bilinear form with $\alpha \circ \beta = 0$ iff $\{\alpha, \beta\}$ is F-dependent.

On the set $G^{\otimes} = F^2 \times F^2 \times F = \{(\alpha, \beta, c) : \alpha, \beta \in F^2, c \in F\}$ define a binary operation

$$(\alpha, \beta, c) \cdot (\alpha', \beta', c') = (\alpha + \alpha', \beta + \beta', c + c' + \beta \circ \alpha'). \tag{1.7}$$

The operation defined in Eq. (1.7) makes $G^{\otimes}$ into a group of order q^5 with center $Z = \{((0,0),(0,0),c) \in G^{\otimes} : c \in F\}$. Note that in the literature the definition of such a binary operation is usually different (the last two coordinates are interchanged). This new point of view has the advantage that the collineations induced by automorphisms of $G^{\otimes}$ may be described using tensor products of matrices in such a way that the composition of such collineations appears greatly simplified. Notation: If the element $(\alpha, \beta, c) \in G^{\otimes}$ has $\alpha = (0,0)$, we usually just write 0 in place of $(0,0)$ and expect the context to make it clear that something like $(0,0,0) \in G^{\otimes}$ really denotes the element $((0,0),(0,0),0) \in G^{\otimes}$.

$G^{\otimes}$ has two important families of (elementary abelian) subgroups of order q^3.

$$\text{For } 0 \neq \gamma \in F^2, \quad \text{put } \mathcal{L}_\gamma = \{(\gamma \otimes \alpha, c) \in G^{\otimes} : \alpha \in F^2, c \in F\}, \tag{1.8}$$

and

$$\text{for } 0 \neq \alpha \in F^2, \quad \text{put } \mathcal{R}_\alpha = \{(\gamma \otimes \alpha, c) \in G^{\otimes} : \gamma \in F^2, c \in F\}. \tag{1.9}$$

Theorem 1.4.1. *Each $\mathcal{L}_\gamma$ and each $\mathcal{R}_\alpha$ are elementary abelian groups of order q^3. And for nonzero $\alpha, \gamma \in F^2$, $\mathcal{L}_\gamma = \mathcal{L}_\alpha$ (resp., $\mathcal{R}_\gamma = \mathcal{R}_\alpha$) iff $\{\alpha, \gamma\}$ is F-dependent, so we may think of the groups $\mathcal{L}_\gamma$ (resp., $\mathcal{R}_\alpha$) as indexed by the points of $PG(1,q)$.*

Proof. This is a routine exercise, but it is convenient to have a reference for the group operations. First consider $\mathcal{L}_\gamma$, where $\gamma = (g_1, g_2)$:

$$\begin{aligned}(\gamma \otimes \alpha, c) \cdot (\gamma \otimes \alpha', c') &= (g_1\alpha, g_2\alpha, c) \cdot (g_1\alpha', g_2\alpha', c') \\ &= (g_1(\alpha + \alpha'), g_2(\alpha + \alpha'), c + c' + g_2\alpha \circ g_1\alpha') \\ &= (\gamma \otimes (\alpha + \alpha'), c + c' + g_1 g_2(\alpha \circ \alpha')),\end{aligned}$$

which is in $\mathcal{L}_\gamma$. And clearly each element of $\mathcal{L}_\gamma$ has order at most 2.

Now consider $\mathcal{R}_\alpha$ with $\gamma = (g_1, g_2)$, $\gamma' = (g_1', g_2')$. Then

$$(\gamma \otimes \alpha, c) \cdot (\gamma' \otimes \alpha, c') = (g_1\alpha, g_2\alpha, c) \cdot (g_1'\alpha, g_2'\alpha, c') = ((\gamma + \gamma') \otimes \alpha, c + c') \in \mathcal{R}_\alpha$$

(since $g_2\alpha \circ g_1'\alpha = 0$). □

Definition 1.4.2. *Define the scalar multiplication $d(\gamma \otimes \alpha, c) = (d\gamma \otimes \alpha, d^2 c) = (\gamma \otimes d\alpha, d^2 c)$ where $d \in F^*$.*

This multiplication makes each $\mathcal{L}_\gamma$ and $\mathcal{R}_\alpha$ into a 3-dimensional vector space over F containing the center Z. Hence there are associated projective planes $\overline{\mathcal{L}}_\gamma$ and $\overline{\mathcal{R}}_\alpha$ isomorphic to $PG(2,q)$.

We pause to consider why we choose this particular scalar multiplication. Suppose we had used $d(\gamma \otimes \alpha, c) = (\gamma \otimes d\alpha, d^i c)$ for some function $d \mapsto d^i$. Then if $\mathcal{L}_\gamma$, $\gamma = (g_1, g_2)$, is a vector space with this scalar multiplication, it must be true that

$$\begin{aligned}
&d[(\gamma \otimes \alpha_1, c_1) \cdot (\gamma \otimes \alpha_2, c_2)] = (\gamma \otimes d\alpha_1, d^i c_1) \cdot (\gamma \otimes d\alpha_2, d^i c_2),\\
&\text{i.e., } d(\gamma \otimes (\alpha_1 + \alpha_2), c_1 + c_2 + g_1 g_2 \alpha_1 \circ \alpha_2)\\
&\qquad\qquad = (\gamma \otimes d(\alpha_1 + \alpha_2), d^i(c_1 + c_2) + g_1 g_2 d^2 \alpha_1 \circ \alpha_2),\\
&\text{i.e., } (\gamma \otimes d(\alpha_1 + \alpha_2), d^i(c_1 + c_2 + g_1 g_2 \alpha_1 \circ \alpha_2))\\
&\qquad\qquad = (\gamma \otimes d(\alpha_1 + \alpha_2), d^i(c_1 + c_2) + g_1 g_2 d^2(\alpha_1 \circ \alpha_2)). \quad (1.10)
\end{aligned}$$

This equality always holds if and only if $d^i = d^2$. So if the same scalar multiplication is used in $\mathcal{L}_\gamma$ as in $\mathcal{R}_\alpha$, clearly we have chosen a reasonable version. The scalar product $d(\gamma \otimes \alpha, c) = (d\gamma \otimes \alpha, dc)$ also makes $\mathcal{R}_\alpha$ into a 3-dimensional vector space over F, but it will turn out a little later not to be as convenient to use this version.

Note. We use the elements of $\tilde{F} = F \cup \{\infty\}$ to index the points of $PG(1, q)$ as follows: $\gamma_\infty = (0, 1)$ and $\gamma_t = (1, t)$ for $t \in F$. So for $(0, 0) \neq \gamma = (a, b)$, we have $\gamma \equiv \gamma_{b/a}$.

Let $\mathcal{C} = \{A_t \equiv \left(\begin{smallmatrix} x_t & y_t \\ 0 & z_t \end{smallmatrix}\right) : t \in F\}$ be a q-clan with $A_0 = \left(\begin{smallmatrix} 0 & 0 \\ 0 & 0 \end{smallmatrix}\right)$. Put $K_t = A_t + A_t^T = y_t P$. Also put $A_\infty = \left(\begin{smallmatrix} 0 & 0 \\ 0 & 0 \end{smallmatrix}\right)$, and $\gamma_{y_\infty} = \gamma_\infty = (0, 1)$. Define then $g(\alpha, t) = \alpha A_t \alpha^T$ for $t \in \tilde{F}$ and $\alpha \in F^2$. It is easy to see that

$$g(\alpha + \beta, t) = g(\alpha, t) + g(\beta, t) + y_t(\alpha \circ \beta). \quad (1.11)$$

For each $t \in \tilde{F}$, there is a subgroup $A(t)$ of $G^\otimes$ defined in the following way:

$$A(t) = \{(\gamma_{y_t} \otimes \alpha, g(\alpha, t)) \in G^\otimes : \alpha \in F^2\} \leq A^*(t) := \mathcal{L}_{\gamma_{y_t}} \leq G^\otimes. \quad (1.12)$$

Using Eq. (1.11) it is easy to check that

$$(\gamma_{y_t} \otimes \alpha_1, g(\alpha_1, t)) \circ (\gamma_{y_t} \otimes \alpha_2, g(\alpha_2, t)) = (\gamma_{y_t} \otimes (\alpha_1 + \alpha_2), g(\alpha_1 + \alpha_2, t)),$$

i.e., $A(t)$ is a subgroup of $\mathcal{L}_{\gamma_{y_t}}$ having order q^2. And $A(t)Z = A^*(t) = \mathcal{L}_{\gamma_{y_t}}$.

Note that in the literature both $g_\alpha(t)$ and $g_t(\alpha)$ have been used to denote the function $g(\alpha, t)$. As we will have occasion to consider this function at different times for fixed α and for fixed t, we have decided not to use these specialized notations to avoid the necessity of having to switch notation in the middle of our argument.

Theorem 1.4.3. *Put $\mathcal{J}(\mathcal{C}) = \{A(t) : t \in \tilde{F}\}$, $\mathcal{J}^*(\mathcal{C}) = \{A^*(t) : t \in \tilde{F}\}$. Then the triple $(G^\otimes, \mathcal{J}(\mathcal{C}), \mathcal{J}^*(\mathcal{C}))$ is a Kantor family, i.e., $\mathcal{J}(\mathcal{C})$ is a 4-gonal family for $G^\otimes$. The associated GQ is denoted $GQ(\mathcal{C})$ and is referred to as a* flock GQ.

Sketch of Proof. For $t, u \in F$, $t \neq u$, we have $y_t \neq y_u$. This leads easily to a proof that $A(t) \cap A^*(u) = \{e\}$ whenever $t, u \in \tilde{F}$, $t \neq u$, which is Kantor's condition K2.

Now suppose s, t, u are distinct elements of $\tilde{F}$ with $\infty \in \{s, t, u\}$. Then it is routine to check that $A(s)A(t) \cap A(u) = \{e\}$ precisely because $A_i - A_j$ is anisotropic, where $\{s, t, u\} = \{\infty, i, j\}$. The hard part is to show that for distinct $t, u, v, \in F$, $A(t)A(u) \cap A(v) = \{e\}$. So suppose

$$(\gamma_{y_t} \otimes \alpha, g(\alpha, t)) \cdot (\gamma_{y_u} \otimes \beta, g(\beta, t)) = (\alpha + \beta, y_t\alpha + y_u\beta, g(\alpha, t) + g(\beta, u) + y_t(\alpha \circ \beta))$$

belongs to $A(v)$, i.e.,

(i) $y_t\alpha + y_u\beta = y_v(\alpha + \beta)$, so that $(y_t + y_v)\alpha = (y_u + y_v)\beta = \gamma$, and

(ii) $g(\alpha, t) + g(\beta, u) + y_t(\alpha \circ \beta) = g(\alpha + \beta, v)$.

From (i) we have $\alpha = (y_t + y_v)^{-1}\gamma$, $\beta = (y_u + y_v)^{-1}\gamma$, so $\alpha \circ \beta = 0$. Putting these values of α and β into (ii) and using Eq. (1.11), we obtain

$$0 = (y_t + y_v)^{-2}(g(\gamma, t) + g(\gamma, v)) + (y_u + y_v)^{-2}(g(\gamma, u) + g(\gamma, v)). \tag{1.13}$$

This says $0 = \gamma B \gamma^T$, where $B = \begin{pmatrix} X & Y \\ 0 & Z \end{pmatrix}$ is the matrix

$$B = \begin{pmatrix} \frac{x_t + x_v}{(y_t+y_v)^2} + \frac{x_u + x_v}{(y_u+y_v)^2} & \frac{y_t+y_v}{(y_t+y_v)^2} + \frac{y_u+y_v}{(y_u+y_v)^2} \\ 0 & \frac{z_t+z_v}{(y_t+y_v)^2} + \frac{z_u+z_v}{(y_u+y_v)^2} \end{pmatrix}.$$

But B is anisotropic, since $1 = tr(XZ/Y^2) = tr(N/D)$, where

$$\begin{aligned} N = x_t z_t (y_v + y_u)^4 + x_u z_u (y_t + y_v)^4 + x_v z_v (y_t + y_u)^4 \\ + (x_t z_u + x_u z_t)(y_v + y_t)^2 (y_v + y_u)^2 \\ + (x_t z_v + x_v z_t)(y_u + y_t)^2 (y_u + y_v)^2 \\ + (x_u z_v + x_v z_u)(y_t + y_u)^2 (y_t + y_v)^2, \end{aligned} \tag{1.14}$$

and

$$D = (y_t + y_u)^2 (y_t + y_v)^2 (y_u + y_v)^2.$$

To see that $1 = tr(N/D)$, note that because $A_t + A_u$, $A_t + A_v$ and $A_u + A_v$ are anisotropic, we have

$$1 = tr \left\{ \frac{(x_t + x_u)(z_t + z_u)}{(y_t + y_u)^2} + \frac{(x_t + x_v)(z_t + z_v)}{(y_t + y_v)^2} + \frac{(x_u + x_v)(z_u + z_v)}{(y_u + y_v)^2} \right\},$$

and this last expression is equal to $tr(N/D)$.

Hence $0 = \gamma B \gamma^T$ implies $\gamma = 0$, which forces $\alpha = \beta = 0$, as desired. Thus $\mathcal{J}(\mathcal{C})$ satisfies K1 also and is a 4-gonal family for $G^{\otimes}$. □

There is an interesting interpretation of the fact that the matrix B above is anisotropic.

Corollary 1.4.4.
Fix $v \in F$. Replace each A_t of $\mathcal{C}$, $v \neq t \in F$, with $A'_t = (y_t + y_v)^{-2}(A_t + A_v)$. Leave $A'_\infty = \left(\begin{smallmatrix}0&0\\0&0\end{smallmatrix}\right)$, and put $A'_v = \left(\begin{smallmatrix}0&0\\0&0\end{smallmatrix}\right)$. Then $\mathcal{C}' = \{A'_t : t \in F\}$ is a q-clan.

The preceding corollary says that if we are given one q-clan (equivalently, one flock), we immediately have a family of $q+1$ q-clans (equivalently, a family of $q+1$ flocks), which might not be pairwise equivalent. This is an analog for characteristic 2 of the concept of "derivation of flocks" introduced by [BLT90] for odd characteristic. We shall meet this concept again when we study the Fundamental Theorem and its extensions.

1.5 Ovals in $\overline{\mathcal{R}_\alpha}$

For each $t \in \tilde{F}$, and for fixed nonzero $\alpha \in F^2$, put

$$A_\alpha(t) = A(t) \cap \mathcal{R}_\alpha = \{(\gamma_{y_t} \otimes d\alpha, g(d\alpha, t)) : d \in F\}. \tag{1.15}$$

Since $g(d\alpha, t) = d^2 g(\alpha, t)$, we have $(\gamma_{y_t} \otimes d\alpha, g(d\alpha, t)) = d(\gamma_{y_t} \otimes \alpha, g(\alpha, t))$, using the scalar multiplication given in Definition 1.4.2 to make $\mathcal{R}_\alpha$ into a 3-dimensional vector space over F. Hence $A_\alpha(t)$ corresponds to a point $\overline{p}_\alpha(t)$ (when $d \in F^*$) in the projective plane $\overline{\mathcal{R}}_\alpha$.

Kantor's property K1, when interpreted for the groups $A_\alpha(t)$ considered as points, says that $\overline{\mathcal{O}}_\alpha = \{\overline{p}_\alpha(t) : t \in \tilde{F}\}$ is an oval (i.e., a set of $q+1$ points of a plane no three on a line) in $\overline{\mathcal{R}}_\alpha$, as proved in Proposition 1.5.2.

Definition 1.5.1. *The set $\{(\gamma_{y_t} \otimes d\alpha, g(d\alpha, t)) : d \in F^*\}$ of nonzero vectors in $A_\alpha(t) = A(t) \cap \mathcal{R}_\alpha$ which correspond to a point of the oval $\overline{\mathcal{O}}_\alpha$ in $\overline{\mathcal{R}}_\alpha$ is called an* o-point.

We will use the notation $\langle(\gamma \otimes \alpha, c)\rangle_1 = \{(\gamma \otimes d\alpha, d^2 c) : d \in F^*\}$ to mean the equivalence class with respect to the scalar multiplication defined in Definition 1.4.2, which represents a point of $\overline{\mathcal{R}}_\alpha$.

Proposition 1.5.2. *$\overline{\mathcal{O}}_\alpha = \{\overline{p}_\alpha(t) : t \in \tilde{F}\}$ is an oval of $\overline{\mathcal{R}}_\alpha$.*

Proof. Take any three distinct points of $\overline{\mathcal{O}}_\alpha$ (the details below hold for points $\overline{p}_\alpha(t)$ with $t \neq \infty$):

$$\overline{p}_\alpha(t) = \langle(\gamma_{y_t} \otimes d\alpha, g(d\alpha, t))\rangle_1 = \langle(d\alpha, y_t d\alpha, g(d\alpha, t))\rangle_1,$$
$$\overline{p}_\alpha(s) = \langle(\gamma_{y_s} \otimes d'\alpha, g(d'\alpha, s))\rangle_1 = \langle(d'\alpha, y_s d'\alpha, g(d'\alpha, s))\rangle_1,$$
$$\overline{p}_\alpha(r) = \langle(\gamma_{y_r} \otimes d''\alpha, g(d''\alpha, r))\rangle_1 = \langle(d''\alpha, y_r d''\alpha, g(d''\alpha, r))\rangle_1,$$

with $t, s, r \in F$, $d, d', d'' \in F$ and $t \neq s \neq r$.

The points in the line spanned by $\overline{p}_\alpha(t)$ and $\overline{p}_\alpha(s)$ have the form:

$$\begin{aligned}\langle\overline{p}_\alpha(t), \overline{p}_\alpha(s)\rangle \\ = \{\langle(\lambda d\alpha + \mu d'\alpha, \lambda y_t d\alpha + \mu y_s d'\alpha, \lambda^2 g(d\alpha, t) + \mu^2 g(d'\alpha, s))\rangle_1 \in \overline{\mathcal{R}}_\alpha : \\ \lambda, \mu \in F, (\lambda, \mu) \neq (0,0)\}.\end{aligned}$$

Suppose now $\overline{p}_\alpha(r) \in \langle \overline{p}_\alpha(t), \overline{p}_\alpha(s) \rangle$, i.e., $\exists \lambda, \mu \in F^*$ such that

$$\begin{cases} d''\alpha = (\lambda d + \mu d')\alpha \\ y_r d''\alpha = (\lambda y_t d + \mu y_s d')\alpha \\ g(d''\alpha, r) = \lambda^2 d^2 g(\alpha, t) + \mu^2 {d'}^2 g(\alpha, s). \end{cases}$$

Since $A(t)A(s) = \{(\tilde{\alpha} + \alpha', y_t\tilde{\alpha} + y_s\alpha', g(\tilde{\alpha}, t) + g(\alpha', s) + y_t\tilde{\alpha} \circ \alpha') : \tilde{\alpha}, \alpha' \in F^2\}$, taking $\tilde{\alpha} = \lambda d\alpha$ and $\alpha' = \mu d'\alpha$, it immediately follows that $\overline{p}_\alpha(r) \in A(r) \cap A(t)A(s)$, which is clearly impossible as Kantor's property K1 holds by hypothesis. Hence any three distinct points of $\overline{\mathcal{O}}_\alpha$ are never collinear and so $\overline{\mathcal{O}}_\alpha$ is an oval of $\overline{\mathcal{R}}_\alpha$. □

1.6 Herd Cover and Herd of Ovals

At this point we pause to make some observations about the scalar multiplication of Definition 1.4.2. These considerations will lead in a natural way to the definitions of *oval cover*, *herd cover* and *profile of a herd cover*.

Since $\overline{\mathcal{R}}_\alpha$ is a projective plane whose points are the equivalence classes $\langle(\gamma \otimes \alpha, c)\rangle_1 = \{(\gamma \otimes d\alpha, d^2c) : d \in F^*\}$ with respect to the scalar multiplication of Definition 1.4.2, it is interesting to look for an isomorphism between $\overline{\mathcal{R}}_\alpha$ and the projective plane $PG(2, q)$ in which the scalar multiplication is defined in the usual way ($\langle \dots \rangle_2$ will denote an equivalence class with respect to the scalar multiplication in the vector space underlying $PG(2, q)$). The following theorem is what we need (observe that the q-clan plays a role in the definition of $g(\alpha, t)$)

Theorem 1.6.1. *A q-clan $\mathcal{C}$ provides a family of $q+1$ ovals $\overline{\mathcal{O}}_\alpha$, one for each $\alpha \in PG(1, q)$, with $\overline{\mathcal{O}}_\alpha = \{\overline{p}_\alpha(t) : t \in \tilde{F}\} \subset \overline{\mathcal{R}}_\alpha$. For each $d \in F^*$ and a fixed $\alpha \neq (0, 0)$ the mapping*

$$\pi_{d\alpha} : \overline{\mathcal{R}}_\alpha \to PG(2, q) : \langle(\gamma \otimes d\alpha, d^2c)\rangle_1 \mapsto \langle(\gamma^{(2)}, d^2c)\rangle_2 \tag{1.16}$$

is an isomorphism of planes.

Moreover, $\pi_{d\alpha}$ maps $\overline{\mathcal{O}}_\alpha$ to the oval $\mathcal{O}_{d\alpha} = \{(\gamma_t, g(d\alpha, t)) : t \in \tilde{F}\}$ provided that $\alpha = (a, b) \neq (0, 0)$.

Here, if $\gamma = (a, b)$, $\gamma^{(2)} := (a^2, b^2)$. We will try to give some justification for the isomorphism $\pi_{d\alpha}$ just introduced. If $d \neq \eta$, unfortunately, $\pi_{d\alpha} \neq \pi_{\eta\alpha}$. By Theorem 1.6.1, corresponding to the oval $\overline{\mathcal{O}}_\alpha = \overline{\mathcal{O}}_{d\alpha}$ in $\overline{\mathcal{R}}_\alpha$ there are $q-1$ projectively equivalent ovals in $PG(2, q)$, namely, $\{\mathcal{O}_{d\alpha} : d \in F^*\}$. There are two approaches that we may take in working with this correspondence. We could associate the oval $\overline{\mathcal{O}}_\alpha$ with the set $\{\mathcal{O}_{d\alpha} : d \in F^*\}$.

Definition 1.6.2. *The set $\{\mathcal{O}_{d\alpha} : d \in F^*\}$is called an* oval cover. *The set $\{\mathcal{O}_{d\alpha} : d \in F^*\}$ is denoted by $[\overline{\mathcal{O}}_\alpha]$.*

We can select otherwise a representative from the oval cover to associate with $\overline{\mathcal{O}}_\alpha$ (this is referred to as *normalizing*). The first approach more accurately reflects the relations in the GQ, but the second approach is easier to deal with computationally. Following the first approach, we get the following definition.

Definition 1.6.3. *A* herd cover, *$\mathcal{H}(\mathcal{C})$, of the q-clan $\mathcal{C}$ is the set of $q+1$ oval covers in $PG(2,q)$ associated to the ovals $\overline{\mathcal{O}}_\alpha$ of the planes $\overline{\mathcal{R}}_\alpha$. That is,*

$$\mathcal{H}(\mathcal{C}) = \{[\overline{\mathcal{O}}_\alpha] \colon \alpha \in PG(1,q)\}. \tag{1.17}$$

This definition is consistent with the terminology introduced in [C02] for a more general setting, provided the indices α are interpreted as points of a conic in $PG(2,q)$. In the current context this distinction is not relevant. As far as the second approach is concerned, normalization is equivalent to choosing, for each $\alpha \in PG(1,q)$, one specific value $d_\alpha \in F^*$ and using only the isomorphisms $\pi_{d_\alpha \alpha}$.

Definition 1.6.4. *When d_α is fixed, the set of $q+1$ ovals $\{\mathcal{O}_{d_\alpha \alpha}\}$ of $PG(2,q)$ obtained letting α vary in $PG(1,q)$ and projecting $\overline{\mathcal{O}}_{d_\alpha \alpha}$ by $\pi_{d_\alpha \alpha}$, is called a* herd of ovals *corresponding to the q-clan $\mathcal{C}$. A herd of ovals is denoted by $\hat{\mathcal{H}}(\mathcal{C})$.*

Definition 1.6.5. *The set $\overline{\mathcal{H}(\mathcal{C})} = \{\overline{\mathcal{O}}_\alpha : \alpha \in PG(1,q)\}$ of $q+1$ ovals of the planes $\overline{\mathcal{R}}_\alpha$ is called the* profile of the herd cover *$\mathcal{H}(\mathcal{C})$, or simply a* profile of the herd.

Observation 1.6.6. The normalization we will use in this book is the one which selects from the oval cover the representative $\mathcal{O}_{d_\alpha \alpha}$, with $\alpha = (1, s^{\frac{1}{2}})$ or $(0,1)$, and $d_\alpha = 1$ for all such α, so that the ovals of the herd will look like $\mathcal{O}_s = \{(1, s, f_s(t)) : t \in F\} \cup \{(0,1,0)\}$ where $f_s(t) = f(t) + (st)^{1/2} + sg(t)$ for suitable maps f, g (as we will see in Theorem 1.7.1, f, g are exactly the q-clan functions defining the matrices of $\mathcal{C}$).

A priori there appears to be no reason to choose one normalization over another. Different normalizations produce different sets of ovals to be called herds of ovals for the same q-clan. From this point of view, a herd of ovals, since it is dependent on the normalization, is not as general as a herd cover. Furthermore, as we shall see in Chapter 3, normalization can have a confounding effect on the computation of automorphism groups. On the other hand, however, we will often adopt the normalization as in Observation 1.6.6 because it is very useful in terms of computations (we will also give some justifications for such a normalization in Chapter 3).

Definition 1.6.7. *An* o-permutation *for $PG(2,q)$ is a permutation polynomial f over $F = GF(q)$ of degree at most $q-2$ with $f(0) = 0$ satisfying the condition that $s(x) = \dfrac{f(x+t)+f(t)}{x}, x \neq 0, s(0) = 0$ is a permutation for each $t \in F$.*

The set of points $\{(1, t, f(t)) \colon t \in F\} \cup \{(0,1,0)\}$ form an oval in $PG(2,q)$ passing through the point $(1,0,0)$ and having nucleus $(0,0,1)$ if and only if f is an o-permutation.

Definition 1.6.8. *If $f(1) = 1$ then an o-permutation f is called an* o-polynomial.

Associated with an o-polynomial are $q-1$ o-permutations, being the nonzero scalar multiples of the o-polynomial. With an o-permutation g_s is associated a unique o-polynomial $\frac{g_s}{g_s(1)}$.

Note that sometimes we will use the notation $\mathcal{D}(f)$ in place of $\mathcal{O}_s$ where f is an o-permutation or an o-polynomial (e.g., Section 3.4), i.e., $\mathcal{D}(f) = \{(1, t, f(t)) : t \in F\} \cup \{(0, 1, 0)\}$ is an oval in $PG(2, q)$. It was first proved in [CPPR96] that if there are o-permutations f and g with $f(0) = g(0) = 0$ for which $f_s^*(t) = f(t) + (st)^{\frac{1}{2}} + sg(t)$ is an o-permutation for all $s \in F$, then

$$\mathcal{C} = \left\{ A_t = \begin{pmatrix} f(t) & t^{\frac{1}{2}} \\ 0 & g(t) \end{pmatrix} : t \in F \right\}$$

is a q-clan (see Theorem 1.7.1) . We point out the obvious statement that for $s \in F$, $f_s^*(t)$ is an o-permutation if and only if $df_s^*(t)$ is an o-permutation for all $d \in F^*$.

1.7 Herds of Ovals from q-Clans

In this section we will use the normalization as in Observation 1.6.6. It was first observed by S. E. Payne and C. Maneri [PM82] (in a specific case) that a q-clan yields a herd of ovals, and the general observation was made more explicit in [Pa85]. The interesting converse, that a herd (i.e., a profile of a herd) yields a q-clan was first given in [CPPR96].

Theorem 1.7.1. *Given a (profile of a) herd of ovals, a q-clan may be constructed. Hence (profiles of) herds and q-clans are equivalent objects.*

Proof. Let $\hat{\mathcal{H}}(\mathcal{C}) = \{\mathcal{O}_s \subset PG(2, q) : s \in \tilde{F}\}$ be a normalized herd of ovals where $f_s(t) = f(t) + (st)^{1/2} + sg(t)$. We will show that $\mathcal{C} = \left\{ A_t \equiv \begin{pmatrix} f(t) & t^{1/2} \\ 0 & g(t) \end{pmatrix} : t \in F \right\}$ is a q-clan. Since $\mathcal{O}_s$ is an oval, any three distinct points of $\mathcal{O}_s$ are never collinear. For arbitrary t, $u \in F$, with $t \neq u$, take the following three points in $\mathcal{O}_s$: $(1, t, f_s(t))$, $(1, u, f_s(u))$, $(0, 1, 0)$. Then $\begin{vmatrix} 1 & t & f_s(t) \\ 1 & u & f_s(u) \\ 0 & 1 & 0 \end{vmatrix} = f_s(u) + f_s(t) \neq 0 \; \forall s \in F$. Hence the following equation in $s^{1/2}$ must have no non-trivial solution:

$$(g(t) + g(u))s + (t + u)^{\frac{1}{2}}(s)^{\frac{1}{2}} + f(t) + f(u) = 0$$

which is equivalent to requiring that the absolute trace of $\frac{(f(t)+f(u))(g(t)+g(u))}{t+u}$ be equal to 1 whenever $t \neq u$. By Observation 1.1.4, that is exactly the condition for the matrix $A_t - A_u$ to be anisotropic. Since it holds for any $t, u \in F$ with $t \neq u$, it immediately follows that $\mathcal{C}$ is a q-clan. Hence a q-clan is constructed from a (profile of a) herd of ovals, and using Theorem 1.6.1 we have that herds and q-clans are equivalent objects. □

1.8 Generalized Quadrangles from q-Clans

Also in this section we will still use the normalization according to Observation 1.6.6. Let $\mathcal{C} = \{A_t \equiv \left(\begin{smallmatrix} xt & yt \\ 0 & zt \end{smallmatrix}\right) : t \in F\}$ be a q-clan normalized so that $A_0 = \left(\begin{smallmatrix} 0 & 0 \\ 0 & 0 \end{smallmatrix}\right)$, and put $A_\infty = \left(\begin{smallmatrix} 0 & 0 \\ 0 & 0 \end{smallmatrix}\right)$. In the previous section we saw that $\mathcal{J}(\mathcal{C})$ is a 4-gonal family for $G^\otimes$. Hence by Kantor's construction (see the Prolegomena) there arises a generalized quadrangle $GQ(\mathcal{C})$ of order (q^2, q) with the following incidence diagram (see Theorem 1.4.3).

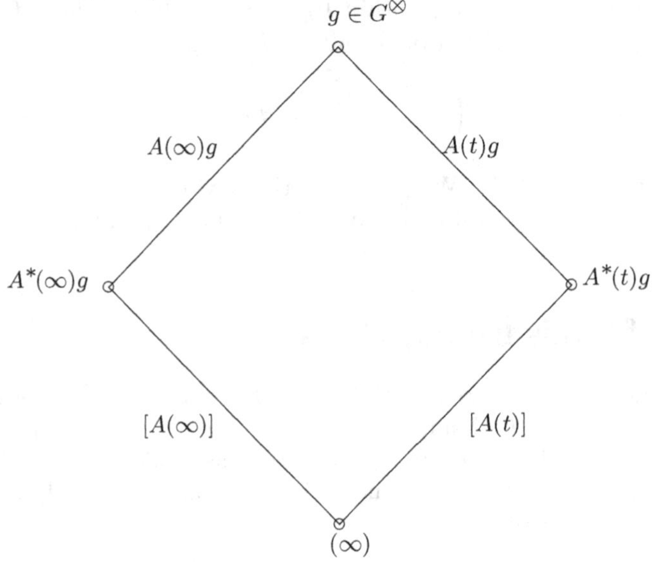

Incidence Diagram for GQ from Kantor Family

Also, as we have just seen, for each $\alpha \in PG(1,q)$ there is an oval $\mathcal{O}_\alpha$ in $PG(2,q)$. We want to observe that there is a subquadrangle $\mathcal{S}_\alpha$ of $GQ(\mathcal{C})$ that is isomorphic to $T_2(\mathcal{O}_\alpha)$ of J. Tits (cf. [De68] or [PT84]). First consider a more general setting.

Theorem 1.8.1 (Payne and Maneri [PM82]). *Let G be a finite group of order s^2t, $1 < s$, $1 < t$. Let $\mathcal{J} = \{A_0, \dots, A_t\}$ be a family of subgroups of G each of order s. Let C be a normal subgroup of G of order t, and put $A_i^* = A_iC$, $0 \le i \le t$. Assume that whenever B_1, B_2, B_3 are three distinct members of $\mathcal{J} \cup \{C\}$, then $B_1B_2 \cap B_3 = \{e\}$. Then if $\mathcal{J}^* = \{A_i^* : 0 \le i \le t\}$, it follows that $(G, \mathcal{J}, \mathcal{J}^*)$ is a Kantor family, and $\mathcal{S}(G, \mathcal{J})$ is a GQ with parameters (s,t). Let G' be a subgroup of G satisfying the following:*

(i) $|G'| = (s')^2t$, $1 < s' < s$;

(ii) $|G' \cap A| = s'$ *for each* $A \in \mathcal{J}$;

(iii) $C \le G'$.

Then $\mathcal{J}' = \{A' = G' \cap A : A \in \mathcal{J}\}$ *is a 4-gonal family for* G', *and* $\mathcal{S}(G', \mathcal{J}')$ *is a subquadrangle of* $\mathcal{S}(G, \mathcal{J})$ *containing the point* (∞) *and having order* (t, t), *i.e.,* $s' = t$.

Sketch of Proof. It is immediate that $\mathcal{J}$ is a 4-gonal family for G and the "tangent space" A_i^* at A_i is $A_i^* = A_i C$. It is also easy to check that $\mathcal{J}'$ is a 4-gonal family for G' with $(\mathcal{J}')^* = \{(A')^* = A'C : A' \in \mathcal{J}'\}$. It needs to be checked that points A^*g of the GQ $\mathcal{S}(G', \mathcal{J}')$ of type (ii) (see the Prolegomena) can be identified with points of $\mathcal{S}(G, \mathcal{J})$ and the lines $A'g$ of $\mathcal{S}(G', \mathcal{J}')$ of type (a) can be identified with lines of $\mathcal{S}(G, \mathcal{J})$. In the first case, if $A^*g = A^*h$ for $A^* \in \mathcal{J}^*$, $g, h \in G'$, then $gh^{-1} \in A^*AC$. Setting $gh^{-1} = ac$ with $a \in A$ and $c \in C$, since $gh^{-1} \in G'$ and $c \in G'$, we conclude that $a \in G'$ and $gh^{-1} \in A'C = (A')^*$, and $(A')^*g = (A')^*h$. In the second case, if $Ag = Ah$ for $A \in \mathcal{J}$ and $g, h \in G'$, then gh^{-1} belongs to A but also to G'. Hence $gh^{-1} \in A'$, implying $A'g = A'h$.

Incidence in $\mathcal{S}(G', \mathcal{J}')$ is easily seen to imply incidence in $\mathcal{S}(G, \mathcal{J})$, so that $\mathcal{S}(G', \mathcal{J}')$ is a subquadrangle of $\mathcal{S}(G, \mathcal{J})$ of order (s', t). By (the point-line dual of) 2.2.1 of [PT84] it must be that $s't \le s$, and by 1.2.3 of [PT84], $s \le t^2$. Hence $s' \le t$. Moreover, right multiplication by elements of C (as in the Prolegomena) induces a group of t symmetries of $\mathcal{S}(G, \mathcal{J})$ about (∞) (collineations fixing each point of $(\infty)^\perp$) leaving $\mathcal{S}(G', \mathcal{J}')$ invariant. Hence (∞) must be regular in both $\mathcal{S}(G, \mathcal{J})$ and $\mathcal{S}(G', \mathcal{J}')$. This forces $s' \ge t$ (cf. 1.3.6 (i) of [PT84]), so that $s' = t$. □

In the present context we have $G = G^\otimes$, $C = Z$, $\mathcal{J} = \{A(t) : t \in \tilde{F}\}$ and $G' = \mathcal{R}_\alpha$ for any choice of $\alpha \in PG(1, q)$. Write $G_\alpha^\otimes = \mathcal{R}_\alpha$, $A_\alpha(t) = A(t) \cap \mathcal{R}_\alpha$, $\mathcal{J}_\alpha = \{A_\alpha(t) : t \in \tilde{F}\}$, $A_\alpha^*(t) = A_\alpha(t)Z = A^*(t) \cap \mathcal{R}_\alpha = \mathcal{L}_{\gamma t} \cap \mathcal{R}_\alpha$.

Define the following subquadrangle of $\mathcal{S}(G, \mathcal{J})$. The *points* of the subGQ $\mathcal{S}_\alpha$ are of the following types:

i) the elements $g = (\gamma \otimes \alpha, c) \in \mathcal{R}_\alpha$;

ii) the cosets $A_\alpha^*(t)g$, $t \in \tilde{F}, g \in \mathcal{R}_\alpha$;

iii) the symbol (∞).

The *lines* of $\mathcal{S}_\alpha$ are of the following types:

a) the cosets $A_\alpha(t)g$, $t \in \tilde{F}, g \in \mathcal{R}_\alpha$;

b) $[A_\alpha(t)]$, $t \in \tilde{F}$.

Incidence is inherited from $\mathcal{S}(G, \mathcal{J})$.

Let Γ be the dual grid with $2(q+1)$ points spanned by the points (∞) and $(0, 0, 0)$. All the subquadrangles $\mathcal{S}_\alpha$ constructed above contain Γ. We know that two subquadrangles of order q in $GQ(\mathcal{C})$ that have a dual grid (with $2(q+1)$ points) in common must have in common just the points and lines of that dual grid (cf. 2.2.2(vi) of [PT84]). Hence $\mathcal{S}_\alpha$ and $\mathcal{S}_\beta$ are identical if $\{\alpha, \beta\}$ is F-dependent, and they meet in Γ otherwise. It follows that we have constructed a family of $q + 1$ subquadrangles $\mathcal{S}_\alpha$ of order q, pairwise intersecting in Γ.

Consider now the projective plane $PG(2,q)$ isomorphic to $\overline{\mathcal{R}}_\alpha$ via the isomorphism π_α and call $\mathcal{O}_\alpha$ the oval in $PG(2,q)$ which is the image under π_α of the oval $\overline{\mathcal{O}}_\alpha$. If we embed $PG(2,q)$ in $PG(3,q)$ as the plane $\pi_\infty : x_0 = 0$, then $\mathcal{O}_\alpha = \{(1,t,g(\alpha,t)) : t \in F\} \cup \{(0,1,0)\}$. Consider the *GQ* $T_2(\mathcal{O}_\alpha)$ constructed from $PG(3,q)$, π_∞, $\mathcal{O}_\alpha$. Recall that the *points* of $T_2(\mathcal{O}_\alpha)$ are of the following types:

i) the points of $PG(3,q)$ not in π_∞;

ii) the planes of $PG(3,q) \setminus \pi_\infty$ which meet π_∞ in a tangent line to $\mathcal{O}_\alpha$;

iii) the symbol (∞).

The *lines* of $T_2(\mathcal{O}_\alpha)$ are of the following types:

a) the lines of $PG(3,q)$ not in π_∞, meeting π_∞ in a point of $\mathcal{O}_\alpha$;

b) the points of π_∞ on $\mathcal{O}_\alpha$.

Incidence is as follows: a point of type i) is incident only with the lines of type a) which contain it; a point of type ii) is incident with all lines of type a) contained in it and with the unique line of type b) on it; the point of type iii) is incident with all the lines of type b) and with no line of type a).

The following is an isomorphism between $\mathcal{S}_\alpha$ and $T_2(\mathcal{O}_\alpha)$:

$$\begin{aligned}
&(\infty) \to (\infty)\\
&(\gamma \otimes \alpha, c) \to (1, \gamma^{(2)}, c)\\
&A_\alpha(t)(\gamma \otimes \alpha, c) \to \langle (0,1,t,g(\alpha,t)), (1,\gamma^{(2)}, c)\rangle\\
&A_\alpha(\infty)(\gamma \otimes \alpha, c) \to \langle (0,0,1,0), (1,\gamma^{(2)}, c)\rangle\\
&A_\alpha(t)^*(\gamma \otimes \alpha, c) \to \langle (0,1,t,g(\alpha,t)), (0,0,0,1), (1,\gamma^{(2)}, c)\rangle\\
&A_\alpha(\infty)^*(\gamma \otimes \alpha, c) \to \langle (0,0,1,0), (0,0,0,1), (1,\gamma^{(2)}, c)\rangle\\
&[A_\alpha(t)] \to (0,1,t,g(\alpha,t))\\
&[A_\alpha(\infty)] \to (0,0,1,0)
\end{aligned}$$

(cf. [De68] or [PT84] and [BOPPR1]).

1.9 Spreads of PG(3, q) Associated with q-Clans

Our main interests in this book are generalized quadrangles, ovals and flocks associated with q-clans. However, starting with a q-clan $\mathcal{C}$ we may construct a spread of $PG(3,q)$, which in turn yields a translation plane of order q^2. There is a considerable body of work on these planes and on planes derived from them. The original construction of a spread from a flock of a quadratic cone (which we now know is equivalent to a q-clan) was given independently by J. A. Thas and by M. Walker [Wa76]. However, we give a more direct construction here clearly equivalent to one given by Gevaert and Johnson [GJ88] and shown by them to be equivalent to the one given by Thas and Walker.

With $P = \left(\begin{smallmatrix} 0 & 1 \\ 1 & 0 \end{smallmatrix}\right)$, as usual, let A and B be any 2×2 matrices over F. It is easy to see that $A \equiv B$ iff $B = A + uP$ for some $u \in F$. A line of $PG(3, q)$ will be given as the row space of a 2×4 matrix with rank 2. For example, if $\mathbf{0}$ (resp., I) is the 2×2 zero (resp., identity) matrix over F, put

$$L_\infty = \langle (\mathbf{0}, I) \rangle . \tag{1.18}$$

Then if $u \in F$ and A is any 2×2 matrix over F, put

$$L_{(A,u)} = \langle (I, A + uP) \rangle . \tag{1.19}$$

Lemma 1.9.1. $\mathcal{R}_A = \{L_\infty\} \cup \{L_{(A,u)} : u \in F\}$ *is a regulus.*

Proof. If $A \equiv B$, then $\mathcal{R}_A = \mathcal{R}_B$, so we may assume $A = \left(\begin{smallmatrix} x & y \\ 0 & z \end{smallmatrix}\right)$. It is clear that $\mathcal{R}_A$ consists of $q + 1$ pairwise skew lines. Fig. 1.1 shows the transversals of $\mathcal{R}_A$ (the horizontal lines for $(0, 0) \neq (a, b) \in F^2$). □

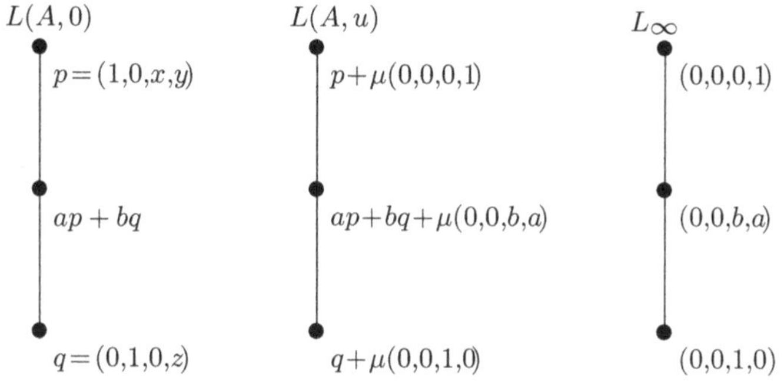

Figure 1.1: $\mathcal{R}_A$ is a regulus

For a line L, let L^* denote the set of points incident with L. And for a matrix A, let $\mathcal{R}_A^*$ denote the set of points incident with a line of the regulus $\mathcal{R}_A$.

Lemma 1.9.2. *Let A, B be two 2×2 matrices over F. Then*

(i) $\langle (I, A) \rangle = \langle (I, B) \rangle$ *iff* $A = B$;

(ii) $\langle (I, A) \rangle \in \mathcal{R}_B$ *iff* $A \equiv B$ *iff* $\mathcal{R}_A = \mathcal{R}_B$;

(iii) $\mathcal{R}_A^* \cap \mathcal{R}_B^* \supseteq L_\infty^*$, *with equality iff $A - B$ is anisotropic.*

Proof. Parts (i) and (ii) are trivial. For part (iii) we may assume (by part (ii)) that $A = \left(\begin{smallmatrix} x1 & y1 \\ 0 & z1 \end{smallmatrix}\right)$ and $B = \left(\begin{smallmatrix} x2 & y2 \\ 0 & z2 \end{smallmatrix}\right)$. Then clearly $\mathcal{R}_A^* \cap \mathcal{R}_B^* \supseteq L_\infty^*$. And the containment is proper iff there exist $u, v \in F$ for which $L_{(A,u)} = \langle (I, A + uP) \rangle$ and $L_{(B,v)} = \langle (I, B + vP) \rangle$ have nontrivial intersection, which holds iff

$$\begin{aligned} 0 &= \det \begin{pmatrix} I & A + uP \\ I & B + vP \end{pmatrix} \\ &= \det \begin{pmatrix} I & A + uP \\ 0 & B - A + (v - u)P \end{pmatrix} \\ &= \det \begin{pmatrix} x_2 - x_1 & y_2 - y_1 + v - u \\ u - v & z_2 - z_1 \end{pmatrix} \\ &= (x_2 - x_1)(z_2 - z_1) + (v - u)(y_2 - y_1) + (v - u)^2 \\ &= \Delta. \end{aligned}$$

If $(x_2 - x_1)(z_2 - z_1) = 0$, we can put $u = 0$ and $v = y_1 - y_2$ to show that $\mathcal{R}_A^* \cap \mathcal{R}_B^* \supseteq L_{(A,u)}^* \cap L_{(B,v)}^* \neq \emptyset$. And it also follows easily in this case that $A - B$ is not anisotropic. So from now on assume that $(x_2 - x_1)(z_2 - z_1) \neq 0$. Then Δ is equal to each of the following:

$$\frac{1}{(z_2 - z_1)} \left[(z_2 - z_1, v - u) \begin{pmatrix} x_2 - x_1 & y_2 - y_1 \\ 0 & z_2 - z_1 \end{pmatrix} \begin{pmatrix} z_2 - z_1 \\ v - u \end{pmatrix} \right], \tag{1.20}$$

and

$$\frac{1}{(x_2 - x_1)} \left[(v - u, x_2 - x_1) \begin{pmatrix} x_2 - x_1 & y_2 - y_1 \\ 0 & z_2 - z_1 \end{pmatrix} \begin{pmatrix} v - u \\ x_2 - x_1 \end{pmatrix} \right]. \tag{1.21}$$

So if $\Delta = 0$, clearly $A - B$ cannot be anisotropic, i.e., if $\mathcal{R}_A^* \cap \mathcal{R}_B^* \neq L_\infty^*$, then $A - B$ is not anisotropic.

Conversely, suppose $\mathcal{R}_A^* \cap \mathcal{R}_B^* = L_\infty^*$. Hence for all $u, v \in F$, $\Delta \neq 0$. We claim that $A - B$ is anisotropic. For suppose $(a, b)(B - A) \left(\begin{smallmatrix} a \\ b \end{smallmatrix}\right) = 0$ with $(0, 0) \neq (a, b)$. If $a \neq 0$, put $u = 0$, $v = (b/a)(z_2 - z_1)$ to contradict $\Delta \neq 0$ in Eq. (1.20). If $b \neq 0$, put $u = 0$ and $v = (a/b)(x_2 - x_1)$ to contradict $\Delta \neq 0$ in Eq. (1.21). □

As an immediate corollary we have the following theorem.

Theorem 1.9.3. *Let $X : t \mapsto x_t$; $Y : t \mapsto y_t$; $Z : t \mapsto z_t$ be three functions from F to F. For $t \in F$ put $A_t = \left(\begin{smallmatrix} x_t & y_t \\ 0 & z_t \end{smallmatrix}\right)$. Then put $\mathcal{C} = \{A_t : t \in F\}$. With the notation as above, $\mathcal{R}_t = \mathcal{R}_{A_t}$ is a regulus for each $t \in F$. Put $\mathcal{S}(\mathcal{C}) = \cup\{\mathcal{R}_t : t \in F\}$. Then $\mathcal{C}$ is a q-clan iff $\mathcal{S}(\mathcal{C})$ is a spread of $PG(3, q)$, in which case X, Y and Z are bijections.* □

Hence with each q-clan $\mathcal{C}$ there is a corresponding line spread $\mathcal{S}(\mathcal{C})$ of $PG(3, q)$, from which a standard construction provides a translation plane $\pi(\mathcal{C})$. It is proved in [GJ88] that two q-clans are equivalent iff their associated translation planes are isomorphic. For our purposes, however, it is sufficient to show how to interpret q-clan equivalence as equivalence of the corresponding spreads. This is taken care of by the following result.

Theorem 1.9.4. *Let $\mathcal{C}$ and $\mathcal{C}'$ be two (not necessarily distinct) q-clans. Then there is a semilinear transformation mapping $\mathcal{S}(\mathcal{C})$ to $\mathcal{S}(\mathcal{C}')$ in such a way that it fixes L_∞ and maps the q special reguli of $\mathcal{S}(\mathcal{C})$ on L_∞ to the q special reguli of $\mathcal{S}(\mathcal{C}')$ if and only if $\mathcal{C} \sim \mathcal{C}'$.*

Proof. First suppose $\mathcal{C} = \{A_t : t \in F\}$ and $\mathcal{C}' = \{A'_t : t \in F\}$ are equivalent. This means that there exist nonzero $\lambda \in F$, $B \in GL(2,q)$, $\sigma \in Aut(F)$, a 2×2 matrix M over F, and a permutation $\pi : t \mapsto \bar{t}$ on the elements of F for which $A'_{\bar{t}} \equiv \lambda B A_t^\sigma B^T + M$ for all $t \in F$, i.e., for each $t \in F$ there is a $u_t \in F$ for which $A'_{\bar{t}} = \lambda B A_t^\sigma B^T + M + u_t P$. Consider the semilinear transformation defined by

$$T : (x_0, x_1, x_2, x_3) \mapsto (x_0^\sigma, x_1^\sigma, x_2^\sigma, x_3^\sigma) \begin{pmatrix} \lambda^{-1}B^{-1} & \lambda^{-1}B^{-1}M \\ 0 & B^T \end{pmatrix}.$$

Clearly T leaves L_∞ invariant, and

$$\begin{aligned} T : L_{(t,u)} &= \langle (I, A_t u P) \rangle \mapsto \langle (\lambda^{-1}B^{-1}, \lambda^{-1}B^{-1}M + (A_t^\sigma + u^\sigma P)B^T) \rangle \\ &= \langle (I, M + \lambda B A_t^\sigma B^T + u^\sigma \lambda B P B^T) \rangle \\ &= \langle (I, \lambda B A_t^\sigma B^T + u^\sigma \lambda (\det(B)) P + M) \rangle \\ &= \langle (I, A'_{\bar{t}} - u_t P + u^\sigma \lambda (\det(B)) P) \rangle. \end{aligned}$$

Hence $T : \mathcal{R}_t \mapsto \mathcal{R}'_{\bar{t}}$ and thus maps $\mathcal{S}(\mathcal{C})$ to $\mathcal{S}(\mathcal{C}')$.

For the converse, suppose there is a semilinear transformation

$$T : (x_0, x_1, x_2, x_3) \mapsto (x_0^\sigma, x_1^\sigma, x_2^\sigma, x_3^\sigma) \begin{pmatrix} C_1 & C_2 \\ C_3 & C_4 \end{pmatrix}$$

(here each C_i is a 2×2 matrix over F) that fixes L_∞ and maps $\mathcal{R}_t$ to $\mathcal{R}'_{\bar{t}}$ for some permutation $\pi : t \mapsto \bar{t}$ of the elements of F. The fact that T fixes L_∞ implies that $C_3 \left(\begin{smallmatrix} 0 & 0 \\ 0 & 0 \end{smallmatrix}\right)$, so C_1 and C_4 are invertible.

It is an easy exercise to show that if A and B are two invertible 2×2 matrices over F, then APB is a scalar multiple of P iff A is a scalar multiple of B^T. Then $T : \langle (I, A_t + uP) \rangle \in \mathcal{R}_t \mapsto \langle (C_1, C_2 + (A_t^\sigma + u^\sigma P) C_4) \rangle = \langle (I, C_1^{-1}C_2 + C_1^{-1} A_t^\sigma C_4 + u^\sigma C_1^{-1} P C_4) \rangle \in \mathcal{R}'_{\bar{t}}$, for all $u \in F$. Since this holds for all $u \in F$, $C_1^{-1} P C_4$ must be a scalar multiple of P, so $C_1^{-1} = \lambda C_4^T$ for some nonzero $\lambda \in F$. Hence

$$T : \langle (I, A_t + uP) \rangle \mapsto \langle (I, \lambda C_4^T A_t^\sigma C_4 + \lambda C_4^T C_2 + \lambda u^\sigma (\det(C_4)) P) \rangle \in \mathcal{R}'_{\bar{t}}.$$

Therefore $A'_{\bar{t}} \equiv \lambda C_4^T A_t^\sigma C_4 + \lambda C_4^T C_2$ for all $t \in F$, implying $\mathcal{C} \sim \mathcal{C}'$. □

Chapter 2

The Fundamental Theorem

The "Fundamental Theorem of q-Clan Geometry" was given that name in [Pa96] because it is the analogue for these geometries of the theorem usually called the Fundamental Theorem of Projective Geometry. In its most basic form (Theorem 2.2.1) it says precisely what are the isomorphisms from one $GQ(\mathcal{C})$ to another $GQ(\mathcal{C}')$ that map $[A(\infty)]$, (0,0,0), and (∞), respectively, of $GQ(\mathcal{C})$ to $[A(\infty)]'$, (0,0,0), and $(\infty)'$, respectively, of $GQ(\mathcal{C}')$. This is really sufficient for some purposes because of the following. If $GQ(\mathcal{C})$ is isomorphic to $GQ(\mathcal{C}')$, then either both are non-classical and then necessarily (∞) maps to $(\infty)'$, or both are classical, in which case we may assume without loss of generality that (∞) maps to $(\infty)'$. And when this occurs, then also without loss of generality we may assume that $(0,0,0)$ maps to $(0,0,0)$. While it is NOT true that without loss of generality $[A(\infty)]$ maps to $[A(\infty)]'$, it is sufficient for our purposes to consider only this case. Then the collineations being considered are all induced by automorphisms of $G^{\otimes}$ of a particularly pleasant form. To determine just when such an automorphism of $G^{\otimes}$ is actually an isomorphism from $GQ(\mathcal{C})$ to $GQ(\mathcal{C}')$ for specific q-clans $\mathcal{C}$ and $\mathcal{C}'$, however, can be a real chore.

When $\mathcal{C}$ is the same as $\mathcal{C}'$, the Fundamental Theorem (F.T.) describes the automorphisms of $GQ(\mathcal{C})$ that fix the line $[A(\infty)]$ and the points (∞) and (0,0,0). (This special case of the F.T. was already given in [PR90], but the opportunity to discover the general result was missed.) By recoordinatizing the GQ so that any line through (∞) may play the role of $[A(\infty)]$, it is possible to extend the F.T. to describe all the automorphisms of $GQ(\mathcal{C})$ that fix the points (∞) and (0,0,0). This is the main goal of this chapter.

In [OP02] an alternative approach to determining the automorphisms of $GQ(\mathcal{C})$ was given that depends on a special "Magic Action" of $P\Gamma L(3,q)$, but an explicit connection between the two approaches was lacking. We have worked out these details in the next chapter.

2.1 Grids and Affine Planes

Since in this section we depend so strongly on the computational setup, we review the notation one more time.

Let $\mathcal{C} = \{A_t \equiv \left(\begin{smallmatrix} xt & yt \\ 0 & zt \end{smallmatrix}\right) : t \in F\}$ be a q-clan normalized so that $A_0 = \left(\begin{smallmatrix} 0 & 0 \\ 0 & 0 \end{smallmatrix}\right)$, and put $A_\infty = \left(\begin{smallmatrix} 0 & 0 \\ 0 & 0 \end{smallmatrix}\right)$. We know that $(\alpha, \beta) \mapsto \alpha \circ \beta = \alpha P \beta^T$ is an alternating, nonsingular bilinear form with $\alpha \circ \beta = 0$ if and only if $\{\alpha, \beta\}$ is F-dependent. Put $g(\alpha, t) = \alpha A_t \alpha^T$, so $g(c\alpha, t) = c^2 g(\alpha, t)$, and $g(\alpha+\beta, t) = g(\alpha, t)+g(\beta, t)+y_t(\alpha\circ\beta)$ for all $\alpha, \beta \in F^2$, $t \in F$. And $g(\alpha, \infty) = 0$ for all $\alpha \in F^2$. The binary operation in $G^\otimes = F^2 \times F^2 \times F$ is given by

$$(\alpha, \beta, c) \cdot (\alpha', \beta', c') = (\alpha + \alpha', \beta + \beta', c + c' + \beta \circ \alpha'), \text{ and}$$
$$(\alpha, \beta, c)^{-1} = (\alpha, \beta, c + \beta \circ \alpha).$$

The 4-gonal family $\mathcal{J}(\mathcal{C})$ for $G^\otimes$ consists of the following $q + 1$ subgroups:

$$\begin{aligned} A(\infty) &= \{(0, \beta, 0) \in G^\otimes : \beta \in F^2\}, \\ A(t) &= \{(\alpha, y_t\alpha, g(\alpha, t)) \in G^\otimes : \alpha \in F^2\}, \ t \in F. \end{aligned}$$

For convenience, we give a listing of when points of $GQ(\mathcal{C})$ are collinear. For points p, q, and for $t \in \tilde{F}$, $p \overset{t}{\sim} q$ means that p and q are elements of $G^\otimes$ belonging to the same right coset of $A(t)$, with a natural adjustment if $q = A^*(t)(\alpha, \beta, c)$:

$$\begin{aligned} (\alpha, \beta, c) &\overset{\infty}{\sim} (\alpha, \beta + \beta', c + \alpha \circ \beta'), \\ (\alpha, \beta, c) &\overset{\infty}{\sim} A^*(\infty)(\alpha, 0, 0), \\ (\alpha, \beta, c) &\overset{t}{\sim} (\alpha + \gamma, \beta + y_t\gamma, c + y_t(\alpha \circ \gamma) + g(\gamma, t)), \\ (\alpha, \beta, c) &\overset{t}{\sim} A^*(t)(0, \beta + y_t\alpha, 0). \end{aligned} \tag{2.1}$$

Now consider the incidence diagram of Fig. 2.1.

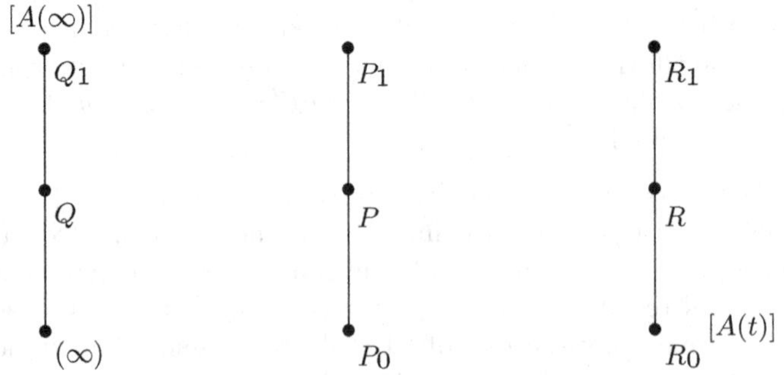

Figure 2.1: Potential Grids on $[A(\infty)]$ and $[A(t)]$, where $P_0 = A^*(t)(0, \beta + \delta + y_t\alpha, 0)$; $R_0 = A^*(t)(0, \beta + y_t\alpha, 0)$; $Q = A^*(\infty)(\alpha + \gamma, 0, 0)$; $Q_1 = A^*(\infty)(\alpha, 0, 0)$; $P_1 = (\alpha, \beta + \delta, c + \alpha \circ \delta)$; $R_1 = (\alpha, \beta, c)$; $R = (\alpha + \gamma, \beta + y_t\gamma, c + y_t\alpha \circ \gamma + g((\gamma, t)))$.

The "missing" point P of Fig. 2.1 exists if and only if $\delta \circ \gamma = 0$, i.e., if and only if $\{\delta, \gamma\}$ is F-dependent. So replace γ with $b\gamma$ and δ with $a\gamma$, where $0 \neq \gamma \in F^2$, to obtain Fig. 2.2.

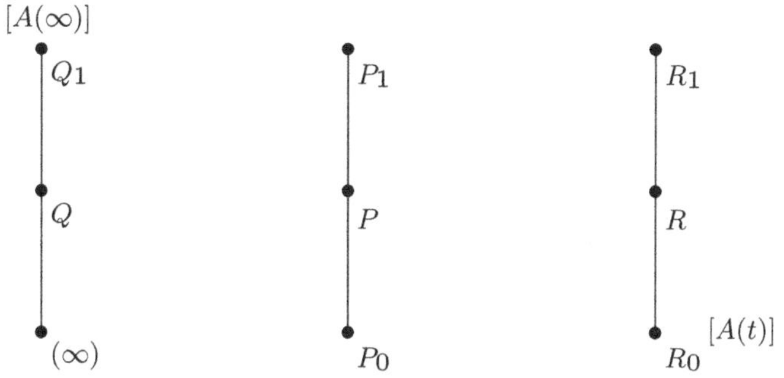

Figure 2.2: $(q+1) \times (q+1)$ Grids on $[A(\infty)]$ and $[A(t)]$, where $P_0 = A^*(t)(0, \beta + a\gamma + y_t\alpha, 0)$; $R_0 = A^*(t)(0, \beta + y_t\alpha, 0)$; $Q = A^*(\infty)(\alpha + b\gamma, 0, 0)$; $Q_1 = A^*(\infty)(\alpha, 0, 0)$; $P = (\alpha + b\gamma, \beta + (a + by_t)\gamma, c + (a + by_t)(\alpha \circ \gamma) + b^2 g(\gamma, t))$; $P_1 = (\alpha, \beta + a\gamma, c + a(\alpha \circ \gamma))$; $R = (\alpha + b\gamma, \beta + y_t\gamma, c + y_t\alpha \circ \gamma + g((\gamma, t)))$; $R_1 = (\alpha, \beta, c)$.

By letting a and b vary independently over the elements of F, we get a $(q+1) \times (q+1)$ grid. Note that the grid lines that meet $[A(\infty)]$ at points other than (∞), meet $[A(\infty)]$ at points of the form $A^*(\infty)(\alpha + b\gamma, 0, 0)$, $b \in F$. This set of points is clearly independent of t. Moreover, if we coordinatize the point $A^*(\infty)(\alpha, 0, 0)$ with $(\alpha, 1) \in PG(2, q)$, the set of points on $[A(\infty)]$ other than (∞) is coordinatized by the line $[\gamma P, \alpha \circ \gamma]^T$ of $PG(2, q)$. So the q^2 points of $[A(\infty)]$ other than (∞) are the affine points of a projective plane $\pi(\infty)$ isomorphic to $PG(2, q)$ with infinite line $[0, 0, 1]^T$ and whose finite lines are the sets of points of $[A(\infty)]$ meeting the lines of a grid (as in Fig. 2.2).

Similarly, the grid lines meeting $[A(t)]$ at points other than (∞) meet it at the points $A^*(t)(0, \beta + a\gamma + y_t\alpha, 0)$, $a \in F$. And the points of $[A(t)]$ different from (∞) are naturally the points of an affine plane. Coordinatize $A^*(t)(0, \beta, 0)$ as $(\beta, 1) \in PG(2, q)$. Then the affine line of points of the form $A^*(t)(0, \beta + a\gamma + y_t\alpha, 0)$, $a \in F$, is coordinatized by the line $[\gamma P, (\beta + y_t\alpha) \circ \gamma]^T$, and the infinite line is $[0, 0, 1]^T$.

We claim that all $(q+1) \times (q+1)$ grids containing (∞) and meeting $[A(t)]$ determine the same affine line. Let $a, b \in F$, $u, t \in F$ with $u \neq t$, and check Fig. 2.3 for a proof.

Aside: It follows from the computations of this section that $GQ(\mathcal{C})$ has *Property (G)* at the point (∞) according to the following definition.

Definition 2.1.1. *A GQ $\mathcal{S}$ with parameters (q^2, q) has* Property (G) *at a point p provided the following holds. Let L_1 and M_1 be distinct lines incident with p. Let L_1, L_2, L_3 and L_4 be distinct lines and M_1, M_2, M_3, M_4 be distinct lines for which $L_i \sim M_j$ whenever $1 \leq i + j \leq 7$. Then $L_4 \sim M_4$.*

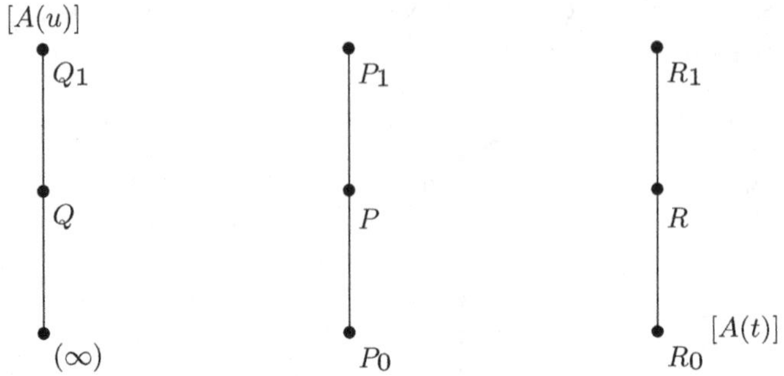

Figure 2.3: $(q+1)\times(q+1)$ Grids on $[A(u)]$ and $[A(t)]$, where
$P_0 = A^*(t)(0,\beta+y_t\alpha+a(y_u+y_t)\gamma,0)$; $R_0 = A^*(t)(0,\beta+y_t\alpha,0)$;
$Q = A^*(u)(0,\beta+y_u\alpha+b(y_t+y_u)\gamma,0)$; $Q_1 = A^*(u)(0,\beta+y_u\alpha,0)$;
$P_1 = (\alpha+a\gamma,\beta+ay_u\gamma,c+ay_u(\alpha\circ\gamma)+a^2g(\gamma,u))$; $R_1 = (\alpha,\beta,c)$;
$P = (\alpha+(a+b)\gamma,\beta+(ay_u+by_t)\gamma,c+(ay_u+by_t)(\alpha\circ\gamma)+a^2g(\gamma,u)+b^2g(\gamma,t))$;
$R = (\alpha+b\gamma,\beta+y_t\gamma,c+by_t(\alpha\circ\gamma)+b^2g(\gamma,t))$.

It was shown in [Pa89] that every flock GQ (i.e., q-clan GQ) has Property (G) at the point (∞), for q odd or even. Remarkably, J. A. Thas [Th99] has shown conversely that having parameters (q^2,q) and having Property (G) at a point completely characterizes flock GQ when q is odd and with an additional hypothesis does so for q even. See Section 8.2 for further remarks on this situation.

2.2 The Fundamental Theorem

With the setup as in the preceding section, assume also that $\mathcal{C}' = \{A'_t : t\in F\}$ is a second (not necessarily distinct) q-clan, with $g'(\alpha,t) = \alpha A'_t\alpha^T$. Let $\theta : GQ(\mathcal{C}) \to GQ(\mathcal{C}')$ be an isomorphism for which $\theta : (\infty) \mapsto (\infty)'$, $\theta : [A(\infty)] \mapsto [A'(\infty)]$, and $\theta : (0,0,0) \mapsto (0,0,0)$. The goal of this section is to determine the form of θ.

Clearly θ is determined as a permutation of the elements of $G^{\otimes}$. And there must be a permutation $t\mapsto \bar{t}$ for which $\theta : A(t) \to A'(\bar{t})$. Moreover, θ must take grids to grids and therefore must map $\pi(\infty)$ to $\pi'(\infty)$ and $\pi(t)$ to $\pi'(\bar{t})$, always mapping infinite points to infinite points and infinite lines to infinite lines. This means that for each $t\in F$, there are $B, D_t \in GL(2,q)$ and $\sigma,\sigma_t \in Aut(F)$ such that

$$\theta : A^*(\infty)(\alpha,0,0) \mapsto (A')^*(\infty)(\alpha^\sigma B,0,0), \tag{2.2}$$

and

$$\theta : A^*(t)(0,\beta,0) \mapsto (A')^*(\bar{t})(0,\beta^{\sigma_t}D_t,0). \tag{2.3}$$

Since $(\alpha,\beta,c) \overset{\infty}{\sim} A^*(\infty)(\alpha,0,0)$, clearly $(\alpha,\beta,c)^\theta \overset{\infty}{\sim} (A')^*(\infty)(\alpha^\sigma B,0,0)$. This implies that as a permutation of the elements of $G^{\otimes}$, θ must have the form

$$\theta : (\alpha,\beta,c) \mapsto (\alpha^\sigma B, (\alpha,\beta,c)^{\theta_2}, (\alpha,\beta,c)^{\theta_3}), \tag{2.4}$$

for some functions θ_2, θ_3 that we proceed to determine.

Apply θ to the collinearity $(\alpha, \beta, c) \overset{t}{\sim} A^*(t)(0, \beta + y_t\alpha, 0)$ to obtain the collinearity $(\alpha^\sigma B, (\alpha, \beta, c)^{\theta 2}, (\alpha, \beta, c)^{\theta 3}) \overset{\bar{t}}{\sim} (A')^*(\bar{t})(0, (\beta + y_t\alpha)^{\sigma t} D_t, 0)$. Hence $(\alpha, \beta, c)^{\theta 2} + y_{\bar{t}}\alpha^\sigma B = (\beta + y_t\alpha)^{\sigma t} D_t$, or

$$(\alpha, \beta, c)^{\theta 2} = (\beta + y_t\alpha)^{\sigma t} D_t + y_{\bar{t}}\alpha^\sigma B \text{ (independent of } c). \tag{2.5}$$

Putting $\alpha = 0$ gives $(0, \beta, c)^{\theta 2} = \beta^{\sigma t} D_t = \beta^{\sigma 0} D_0$ for all $t \in F$. Put $\beta = (1, 0)$, then $\beta = (0, 1)$, to get $D_t = D_0 = D$ for all $t \in F$. And D is invertible, so $\sigma_t = \sigma_0$ for all $t \in F$. So now Eq. (2.5) says

$$(\alpha, \beta, c)^{\theta 2} = (\beta + y_t\alpha)^{\sigma 0} D + y_{\bar{t}}\alpha^\sigma B, \text{ for all } t \in F; \alpha, \beta \in F^2. \tag{2.6}$$

From $(\beta + y_t\alpha)^{\sigma 0} D + y_{\bar{t}}\alpha^\sigma B = (\beta + y_s\alpha)^{\sigma 0} D + y_{\bar{s}}\alpha^\sigma B$, we obtain

$$(y_t + y_s)^{\sigma 0}\alpha^{\sigma 0} D = (y_{\bar{t}} + y_{\bar{s}})\alpha^\sigma B. \tag{2.7}$$

Put $\alpha = (1, 0)$ and then $\alpha = (0, 1)$ in Eq. (2.7) to obtain

$$(y_t + y_s)^{\sigma 0} D = (y_{\bar{t}} + y_{\bar{s}}) B. \tag{2.8}$$

Then put Eq. (2.8) into Eq. (2.7):

$$\alpha^{\sigma 0}[(y_t + y_s)^{\sigma 0} D] = \alpha^\sigma[(y_{\bar{t}} + y_{\bar{s}}) B] = \alpha^\sigma[(y_t + y_s)^{\sigma 0} D].$$

As D is invertible, it follows that

$$\sigma_0 = \sigma. \tag{2.9}$$

Take determinants in Eq (2.8): $(y_t + y_s)^{2\sigma}|D| = (y_{\bar{t}} + y_{\bar{s}})^2|B|$. So again from Eq. (2.8),

$$D = \left(\frac{y_{\bar{t}} + y_{\bar{s}}}{(y_t + y_s)^\sigma}\right) B = |D \cdot B^{-1}|^{1/2} \cdot B. \tag{2.10}$$

Put $T = |DB^{-1}|^{1/2}$, so $D = TB$, and $s = 0$ gives

$$y_{\bar{t}} = y_t^\sigma T + y_{\bar{0}}. \tag{2.11}$$

So now $(\alpha, \beta, c)^{\theta 2} \overset{Eq. (2.6)}{=} (\beta + y_t\alpha)^\sigma D + y_{\bar{t}}\alpha^\sigma B = (\beta + y_t\alpha)^\sigma TB + (y_t^\sigma T + y_{\bar{0}})\alpha^\sigma B = T\beta^\sigma B + y_{\bar{0}}\alpha^\sigma B$, which implies

$$(\alpha, \beta, c)^{\theta 2} = (T\beta^\sigma + y_{\bar{0}}\alpha^\sigma) B. \tag{2.12}$$

At this point we have

$$(\alpha, \beta, c)^\theta = (\alpha^\sigma B, (T\beta^\sigma + y_{\bar{0}}\alpha^\sigma) B, (\alpha, \beta, c)^{\theta 3}). \tag{2.13}$$

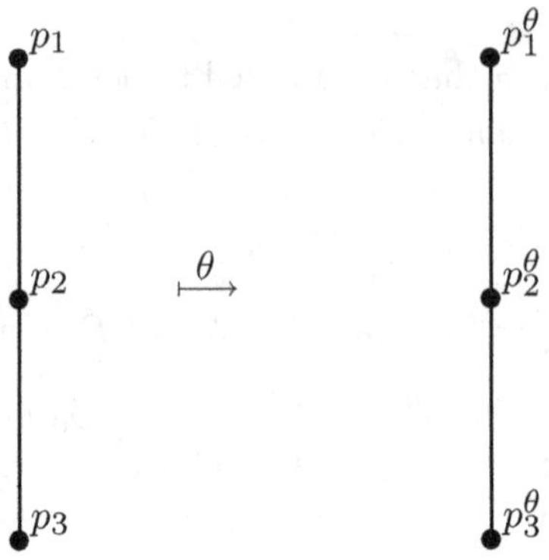

Figure 2.4: $p_1 = (\alpha, \beta, c)$, $p_2 = (0, \beta + y_t\alpha, c + g(\alpha, t))$, $p_3 = A^*(t)(0, \beta + y_t\alpha, 0)$ and $p_1^\theta = (\alpha^\sigma G, (T\beta^\sigma + y_{\bar{0}}\alpha^\sigma)B, (\alpha, \beta, c)^{\theta 3})$, $p_2^\theta = (0, T(\beta + y_t\alpha)^\sigma B, (0, \beta + y_t\alpha, c + g(\alpha, t))^{\theta 3})$, $p_3^\theta = (A')^*(\bar{t})(0, T(\beta + y_t\alpha)^\sigma B, 0)$.

Now use the fact that p_1^θ and p_2^θ must be in the same coset of $A'(\bar{t})$ (and $(p_2^\theta)^{-1} = p_2^\theta$ as elements of $G^{\otimes}$) to obtain that

$$\begin{aligned}&(\alpha^\sigma B, (T\beta^\sigma + y_{\bar{0}}\alpha^\sigma)B, (\alpha, \beta, c)^{\theta 3}) \cdot (0, T(\beta + y_t\alpha)^\sigma B, (0, \beta + y_t\alpha, c + g(\alpha, t))^{\theta 3})\\ &= (\alpha^\sigma B, y_{\bar{t}}\alpha^\sigma B, (\alpha, \beta, c)^{\theta 3} + (0, \beta + y_t\alpha, c + g(\alpha, t))^{\theta 3})\end{aligned}$$

must be in $A'(\bar{t})$. This holds iff

$$(\alpha, \beta, c)^{\theta 3} = (0, \beta + y_t\alpha, c + g(\alpha, t))^{\theta 3} + g'(\alpha^\sigma B, \bar{t}). \tag{2.14}$$

Put $t = 0$ in Eq. (2.14) to obtain

$$(\alpha, \beta, c)^{\theta 3} = (0, \beta, c)^{\theta 3} + g'(\alpha^\sigma B, \bar{0}). \tag{2.15}$$

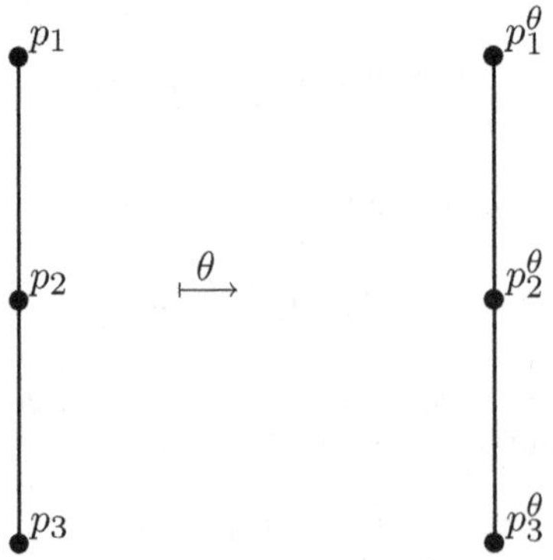

Figure 2.5: $p_1 = (\alpha, \beta, c)$, $p_2 = (\alpha, 0, c + \alpha \circ \beta)$, $p_3 = A^*(\infty)(\alpha, 0, 0)$ and $p_1^\theta = (\alpha^\sigma B, (T\beta^\sigma + y_{\bar{0}}\alpha^\sigma)B, (\alpha, \beta, c)^{\theta 3})$, $p_2^\theta = (\alpha^\sigma B, y_{\bar{0}}\alpha^\sigma B, (\alpha, 0, c + \alpha \circ \beta)^{\theta 3})$, $p_3^\theta = (A')^*(\infty)(\alpha^\sigma B, 0, 0)$.

Now use $(p_1^\theta)(p_2^\theta)^{-1} \in A'(\infty)$ and $(p_2^\theta)^{-1} = (p_2^\theta)$:

$$\begin{aligned}(\alpha^\sigma B, (T\beta^\sigma + y_{\bar{0}}\alpha^\sigma)B, (\alpha,\beta,c)^{\theta 3}) \cdot (\alpha^\sigma B, y_{\bar{0}}\alpha^\sigma B, (\alpha, 0, c+\alpha\circ\beta)^{\theta 3})\\ = (0, T\beta^\sigma B, (\alpha,\beta,c)^{\theta 3} + (\alpha, 0, c+\alpha\circ\beta)^{\theta 3} + (T\beta^\sigma B)\circ(\alpha^\sigma B)) \in A'(\infty)\\ \text{i.e., } (\alpha,\beta,c)^{\theta 3} = (\alpha, 0, c+\alpha\circ\beta)^{\theta 3} + T\beta^\sigma(BPB^T)(\alpha^\sigma)^T. \quad (2.16)\end{aligned}$$

Put $\alpha = 0$ to obtain

$$(0,\beta,c)^{\theta 3} = (0,0,c)^{\theta 3}, \text{ independent of } \beta. \tag{2.17}$$

So Eq. (2.15) becomes:

$$(\alpha,\beta,c)^{\theta 3} = (0,0,c)^{\theta 3} + g'(\alpha^\sigma B, \bar{0}). \tag{2.18}$$

Since $BPB^T = \det(B)\cdot P$, $T\beta^\sigma(BPB^T)(\alpha^\sigma)^T = T\cdot\det(B)(\alpha\circ\beta)^\sigma$. And of course $(0,0,0)^\theta = (0,0,0)$ implies $(0,0,0)^{\theta 3} = 0$. Also,

$$\begin{aligned}(0,0,c)^{\theta 3} + g'(\alpha^\sigma B, \bar{0}) \overset{2.18}{=} (\alpha,\beta,c)^{\theta 3} \overset{2.16}{=} (\alpha, 0, c+\alpha\circ\beta)^{\theta 3} + T\det(B)(\alpha\circ\beta)^\sigma\\ \overset{2.18}{=} (0,0,c+\alpha\circ\beta)^{\theta 3} + g'(\alpha^\sigma B, \bar{0}) + T\det(B)(\alpha\circ\beta)^\sigma\\ \text{i.e., } (0,0,c)^{\theta 3} = (0,0,c+\alpha\circ\beta)^{\theta 3} + T\det(B)(\alpha\circ\beta)^\sigma. \quad (2.19)\end{aligned}$$

Put $c = 0$ in Eq. (2.19): $(0,0,\alpha\circ\beta)^{\theta 3} = T\det(B)(\alpha\circ\beta)^\sigma$, i.e.,

$$(0,0,c)^{\theta 3} = T\det(B)c^\sigma. \tag{2.20}$$

Now put $\Delta = \det(B)$ and use Eqs. (2.20) and (2.18) in Eq. (2.13):

$$(\alpha,\beta,c)^\theta = (\alpha^\sigma B, (T\beta^\sigma + y_{\bar{0}}\alpha^\sigma)B, T\Delta c^\sigma + g'(\alpha^\sigma B, \bar{0})). \tag{2.21}$$

The scalar T is not completely determined yet, but for any $T \in F$ used in Eq. (2.21) the θ described is an automorphism of $G^\otimes$. To yield an isomorphism from $GQ(\mathcal{C})$ to $GQ(\mathcal{C}')$, θ must map $A(t)$ to $A'(\bar{t})$ for each $t \in F$. This says $(\alpha, y_t\alpha, g(\alpha,t))^\theta = (\alpha^\sigma B, (Ty_t^\sigma\alpha^\sigma + y_{\bar{0}}\alpha^\sigma)B, T\Delta g(\alpha,t)^\sigma + g'(\alpha^\sigma B, \bar{0}))$ must be in $A'(\bar{t})$. The middle term is O.K. by Eq. (2.11). So the condition that θ map $A(t)$ to $A'(\bar{t})$ is precisely that $T\Delta g(\alpha,t)^\sigma + g'(\alpha^\sigma B, \bar{0}) = g'(\alpha^\sigma B, \bar{t})$, which holds iff

$$\alpha^\sigma B[T\Delta B^{-1}A_t^\sigma B^{-T} + A'_{\bar{0}} + A'_{\bar{t}}](\alpha^\sigma B)^T = 0$$

(for all $\alpha \in F^2$, $t \in F$), which holds iff

$$A'_{\bar{t}} \equiv T\Delta B^{-1}A_t^\sigma B^{-T} + A'_{\bar{0}} \text{ for all } \theta \in F. \tag{2.22}$$

Put $\lambda = T\Delta$, i.e., $T = \lambda/\Delta$. Then the description of θ in Eq. (2.21) may be written

$$((\alpha,\beta),c)^\theta = ((\alpha^\sigma,\beta^\sigma)\begin{pmatrix} 1 & y_{\bar{0}} \\ 0 & \lambda/\Delta \end{pmatrix} \otimes B, \lambda c^\sigma + g'(\alpha^\sigma B, \bar{0}). \tag{2.23}$$

And we can now complete the proof of the following fundamental theorem.

Theorem 2.2.1 (The Fundamental Theorem). *Let* $\mathcal{C} = \{A_t \equiv \left(\begin{smallmatrix} xt & yt \\ 0 & zt \end{smallmatrix}\right) : t \in F\}$ *and* $\mathcal{C}' = \{A'_t \equiv \left(\begin{smallmatrix} x'_t & y'_t \\ 0 & z'_t \end{smallmatrix}\right) : t \in F\}$ *be two (not necessarily distinct) q-clans normalized so that* $A_0 = \left(\begin{smallmatrix} 0 & 0 \\ 0 & 0 \end{smallmatrix}\right) = A'_0$. *Then the following are equivalent:*

(i) $\mathcal{C} \sim \mathcal{C}'$.

(ii) *The flocks* $\mathcal{F}(\mathcal{C})$ *and* $\mathcal{F}(\mathcal{C})'$ *are projectively equivalent.*

(iii) $GQ(\mathcal{C})$ *and* $GQ(\mathcal{C})'$ *are isomorphic by an isomorphism mapping* $(\infty) \mapsto (\infty)$, $[A(\infty)] \mapsto [A'(\infty)]$, *and* $(0,0,0) \mapsto (0,0,0)$.

(iv) *The associated spreads* $\mathcal{S}(\mathcal{C})$ *and* $\mathcal{S}(\mathcal{C})'$ *are equivalent by a semilinear transformation leaving* L_∞ *fixed and mapping the special reguli of* $\mathcal{S}(\mathcal{C})$ *to the special reguli of* $\mathcal{S}(\mathcal{C})'$.

Moreover, these four conditions hold in a canonical way for each 4-tuple $(\lambda, B, \sigma, \pi)$ *with* $0 \neq \lambda \in F$, $B \in GL(2,q)$, $\sigma \in Aut(F)$, $\pi\colon t \mapsto \bar{t}$ *a permutation on* F *for which the following condition* (v) *holds (put* $\Delta = \det(B)$*):*

(v) $A'_{\bar{t}} \equiv \lambda B^{-1} A_t^\sigma B^{-T} + A'_{\bar{0}}$ *for all* $t \in F$.

(vi) *The corresponding isomorphism* $\theta : GQ(\mathcal{C}) \mapsto GQ(\mathcal{C}')$ *is given by*

$$\theta = \theta(\lambda, B, \sigma, \pi) : ((\alpha,\beta),c) \mapsto ((\alpha^\sigma, \beta^\sigma)\begin{pmatrix} 1 & y_{\bar{0}} \\ 0 & \lambda/\Delta \end{pmatrix} \otimes B, \lambda c^\sigma + g'(\alpha^\sigma B, \bar{0})).$$

(vii) *The corresponding isomorphism between the spreads* $\mathcal{S}(\mathcal{C})$ *and* $\mathcal{S}(\mathcal{C}')$ *is given by*

$$T\colon \mathcal{S}(\mathcal{C}) \mapsto \mathcal{S}(\mathcal{C}')\colon (x_0, x_1, x_2, x_3) \mapsto (x_0^\sigma, x_1^\sigma, x_2^\sigma, x_3^\sigma)\begin{pmatrix} \lambda^{-1}B & \lambda^{-1}BA'_{\bar{0}} \\ 0 & B^{-T} \end{pmatrix}.$$

(viii) *If* $B^{-1} = \left(\begin{smallmatrix} a & b \\ c & d \end{smallmatrix}\right)$, *the corresponding isomorphism* $T_\theta : \mathcal{F}(\mathcal{C}) \to \mathcal{F}(\mathcal{C}')$ *is defined for planes of the flock by*

$$T_\theta\colon \begin{bmatrix} x_t \\ y_t \\ z_t \\ 1 \end{bmatrix} \mapsto \begin{bmatrix} x'_{\bar{t}} \\ y'_{\bar{t}} \\ z'_{\bar{t}} \\ 1 \end{bmatrix} = \begin{pmatrix} \lambda a^2 & \lambda ab & \lambda b^2 & x'_{\bar{0}} \\ 0 & \lambda(ad+bc) & 0 & y'_{\bar{0}} \\ \lambda c^2 & \lambda cd & \lambda d^2 & z'_{\bar{0}} \\ 0 & 0 & 0 & 1 \end{pmatrix}\begin{bmatrix} x_t^\sigma \\ y_t^\sigma \\ z_t^\sigma \\ 1 \end{bmatrix}.$$

2.3 Aut($\mathbf{G}^\otimes$)

Definition 2.3.1. *Let* $\mathcal{G}$ *be the full group of collineations of* $GQ(\mathcal{C})$.

Definition 2.3.2. *Let* $\mathcal{G}_0$ *be the subgroup of* $\mathcal{G}$ *fixing the points* $(0,0,0)$ *and* (∞).

Definition 2.3.3. *Let* $\mathcal{H}$ *be the subgroup of* $\mathcal{G}$ *fixing the points* $(0,0,0)$ *and* (∞) *and the line* $[A(\infty)]$.

It will follow later as a consequence of the Fundamental Theorem that the collineation group of $GQ(\mathcal{C})$ fixing (∞) and $(0,0,0)$ may be viewed as a subgroup of $Aut(G^{\otimes})$ whose elements have an especially nice form. So in this section we consider such automorphisms of $G^{\otimes}$.

Let $\sigma \in Aut(F)$, $0 \neq x \in F$, and let H and S be 4×4 matrices over F. Define $\theta(\sigma, x, H, S) : G^{\otimes} \to G^{\otimes}$ by

$$\theta(\sigma, x, H, S) : ((\alpha, \beta), c) \mapsto ((\alpha^{\sigma}, \beta^{\sigma})H, xc^{\sigma} + (\alpha^{\sigma}, \beta^{\sigma})S(\alpha^{\sigma}, \beta^{\sigma})^T). \tag{2.24}$$

A routine computation shows that

$$\begin{aligned}&\theta(\sigma_1, x_1, H_1, S_1) \circ \theta(\sigma_2, x_2, H_2, S_2)\\ &\qquad = \theta(\sigma_1 \circ \sigma_2, x_1^{\sigma 2} x_2, H_1^{\sigma 2} H_2, x_2 S_1^{\sigma 2} + H_1^{\sigma 2} S_2 (H_1^{\sigma 2})^T).\end{aligned} \tag{2.25}$$

For a given $\theta = \theta(\sigma, x, H, S)$, we may write H and S in block form, with 2×2 blocks. WLOG we may assume that S is in block upper triangular form (since S is only used to determine a quadratic form in four variables). Say $H = \left(\begin{smallmatrix} A & K \\ B & M \end{smallmatrix}\right)$, $S = \left(\begin{smallmatrix} C & D \\ 0 & E \end{smallmatrix}\right)$. Clearly C (resp., E) may be replaced with any matrix $C' \equiv C$ (resp., $E' \equiv E$), so we usually take C and E to be upper triangular. Then a straightforward computation expressing the condition that θ preserve the group operation shows that θ is an automorphism of $G^{\otimes}$ iff H is nonsingular and

$$\begin{array}{ll} \text{(i)} & C + C^T = KPA^T, \\ \text{(ii)} & D = KPB^T, \\ \text{(iii)} & D = APM^T + xP, \\ \text{(iv)} & E + E^T = MPB^T. \end{array} \tag{2.26}$$

For each automorphism of $G^{\otimes}$ of interest here there will be matrices $\overline{A} = \left(\begin{smallmatrix} d & b \\ c & a \end{smallmatrix}\right)$ and $\overline{B}$, both in $GL(2,q)$, for which $H = \overline{A} \otimes \overline{B}$. Put $\lambda = \det(\overline{A})$, $\Delta = \det(\overline{B})$. Then we may rewrite Eq. (2.24) as

$$\begin{aligned}\theta = \ & \theta\left(\sigma, x, \overline{A} \otimes \overline{B}, \begin{pmatrix} C & D \\ 0 & E \end{pmatrix}\right) : ((\alpha, \beta), c) \mapsto \\ & ((\alpha^{\sigma}, \beta^{\sigma})(\overline{A} \otimes \overline{B}), xc^{\sigma} + \alpha^{\sigma} C (\alpha^{\sigma})^T + \alpha^{\sigma} D (\beta^{\sigma})^T + \beta^{\sigma} E (\beta^{\sigma})^T).\end{aligned} \tag{2.27}$$

And the conditions of Eq. (2.26) are equivalent to:

$$\begin{array}{ll} \text{(i)} & C + C^t = \Delta bdP, \\ \text{(ii)} & D = \Delta bcP, \\ \text{(iii)} & x = \lambda\Delta = \sqrt{\det(H)}, \\ \text{(iv)} & E + E^T = \Delta acP. \end{array} \tag{2.28}$$

This means that the symbols x and D may be suppressed in the notation for θ. So temporarily we write

$$\begin{aligned}\theta \ & = \theta(\sigma, \overline{A} \otimes \overline{B}, C, E) : ((\alpha, \beta), t) \mapsto \\ & ((\alpha^{\sigma}, \beta^{\sigma})(\overline{A} \otimes \overline{B}), \lambda\Delta t^{\sigma} + \alpha^{\sigma} C (\alpha^{\sigma})^T + \Delta bc(\alpha^{\sigma} \circ \beta^{\sigma}) + \beta^{\sigma} E (\beta^{\sigma})^T),\end{aligned} \tag{2.29}$$

where $\lambda = \det \overline{A}$, $\Delta = \det(\overline{B})$, $C + C^T \Delta bdP$, and $E + E^T = \Delta acP$. Writing H in the form $\overline{A} \otimes \overline{B}$ makes it easy to track the action of $\theta(\sigma, \overline{A} \otimes \overline{B}, C, E)$ on the subgroups $\mathcal{L}_\gamma$ and $\mathcal{R}_\alpha$ of $G^\otimes$.

$$\begin{aligned} &\text{(i)} \quad \theta(\sigma, \overline{A} \otimes \overline{B}, C, E) : \mathcal{L}_\gamma \to \mathcal{L}_{\gamma^\sigma \overline{A}}; \\ &\text{(ii)} \quad \theta(\sigma, \overline{A} \otimes \overline{B}, C, E) : \mathcal{R}_\alpha \to \mathcal{R}_{\alpha^\sigma \overline{B}}. \end{aligned} \tag{2.30}$$

Using elements of $\tilde{F}$ to label points of $PG(1,q)$ as we did earlier (i.e., $t \leftrightarrow (1,t)$), if $\overline{A}\left(\begin{smallmatrix} a4 & a2 \\ a3 & a1 \end{smallmatrix}\right)$ and $\overline{B}\left(\begin{smallmatrix} b4 & b2 \\ b3 & b1 \end{smallmatrix}\right)$, then we have

$$\begin{aligned} &\text{(i)} \quad \theta(\sigma, \overline{A} \otimes \overline{B}, C, E) : \mathcal{L}_{\gamma t} \to \mathcal{L}_{\gamma \bar{t}}, \text{ where } \bar{t} = \frac{a_1 t^\sigma + a_2}{a_3 t^\sigma + a_4}, \quad t \in \tilde{F}; \\ &\text{(ii)} \quad \theta(\sigma, \overline{A} \otimes \overline{B}, C, E) : \mathcal{R}_{\alpha t} \to \mathcal{R}_{\alpha \bar{t}}, \text{ where } \bar{t} = \frac{b_1 t^\sigma + b_2}{b_3 t^\sigma + b_4}, \quad t \in \tilde{F}. \end{aligned} \tag{2.31}$$

Let $\theta = \theta(\sigma, \overline{A} \otimes \overline{B}, C, E)$ be an automorphism of $G^\otimes$ as given in Eq. (2.29). We want to determine *necessary* conditions on C and E for θ to map the members of the 4-gonal family $\mathcal{J}(\mathcal{C})$ to the members of $\mathcal{J}(\mathcal{C}')$. We know by Eq. (2.31) that $\theta : \mathcal{L}_{\gamma y_t} \to \mathcal{L}_{\gamma y_{\bar{t}}}$, where $y_{\bar{t}} = \frac{a1 y_t^\sigma + a2}{a3 y_t^\sigma + a4}$.

Since $A(t) \leq \mathcal{L}_{\gamma y_t}$ and $A'(t) \leq \mathcal{L}_{\gamma'_{y_t}}$, if θ does map members of $\mathcal{J}(\mathcal{C})$ to those of $\mathcal{J}(\mathcal{C}')$, there must be a permutation $\pi : \tilde{F} \to \tilde{F} : t \mapsto \bar{t}$ satisfying

$$y'_{\bar{t}} = \frac{a_1 y_t^\sigma + a_2}{a_3 y_t^\sigma + a_4} \quad (\text{where } \theta : A(t) \to A'(\bar{t})). \tag{2.32}$$

Since $t \mapsto y_t$ is a permutation of the elements of F, the members of $\mathcal{C}$ could be indexed so that $t \mapsto y_t$ is any given permutation, say $y_t = t$. In the studies made of the known examples, there has always been some multiplicative permutation τ (so fixing 0 and 1 and by convention here also fixing ∞) such that for all $A_t = \left(\begin{smallmatrix} xt & yt \\ 0 & zt \end{smallmatrix}\right) \in \mathcal{C}$, we have $y_t = t^{\tau^{-1}}$. In this case we say that $\mathcal{C}$ is τ^{-1}-*normalized*, with τ^{-1} being multiplicative. Then Eq. (2.32) becomes

$$\bar{t} = \left(\frac{a_1 t^{\sigma/\tau} + a_2}{a_3 t^{\sigma/\tau} + a_4} \right)^\tau. \tag{2.33}$$

First consider $t = 0$, so $\bar{t} = \bar{0} = (a_2/a_4)^\tau$. We now determine the image of $A(0)$. Here

$$\begin{aligned} &\theta(\sigma, \overline{A} \otimes \overline{B}, C, E) : \\ &((1,0) \otimes \alpha, 0) \mapsto \ ((1,0)\overline{A} \otimes \alpha^\sigma \overline{B}, (\alpha^\sigma, 0) \begin{pmatrix} C & \Delta a_2 a_3 P \\ 0 & E \end{pmatrix} (\alpha^\sigma, 0)^T) \\ &\qquad = ((a_4, a_2) \otimes \alpha^\sigma \overline{B}, \alpha^\sigma C (\alpha^\sigma)^T), \end{aligned}$$

which must be in $A'((a_2/a_4)^\tau)$. If $a_4 \neq 0$, put $s = a_2/a_4 \in F$, so $(a_4, a_2) \equiv (1, s) = \gamma_s$. And $((a_4, a_2) \otimes \alpha^\sigma \overline{B}, \alpha^\sigma C(\alpha^\sigma)^T) = ((1,s) \otimes a_4 \alpha^\sigma \overline{B}, \alpha^\sigma C(\alpha^\sigma)^T)$ is in $\mathcal{L}_{\gamma_s}$, which contains $A'(s^\tau)$. Then $((1,s)\otimes a_4\alpha^\sigma\overline{B}, \alpha^\sigma C(\alpha^\sigma)^T) = (\gamma_s \otimes a_4\alpha^\sigma\overline{B}, \alpha^\sigma C(\alpha^\sigma)^T)$ is in $A'(s^\tau)$ if and only if $\alpha^\sigma C(\alpha^\sigma)^T = g'_{s^\tau}(a_4\alpha^\sigma\overline{B}) = a_4^2\alpha^\sigma\overline{B}A'_{s^\tau}\overline{B}^T(\alpha^\sigma)^T$. This holds for all $\alpha \in F^2$ if and only if $C \equiv a_4^2\overline{B}A'_{s^\tau}\overline{B}^T$. Now suppose $a_4 = 0$, so $s = \infty$. Then $((0,a_2)\otimes\alpha^\sigma\overline{B}, \alpha^\sigma C(\alpha^\sigma)^T) \in A'(\infty) = A'(\infty^\tau)$ if and only if $\alpha^\sigma C(\alpha^\sigma)^T = 0$. This holds for all $\alpha \in F^2$ if and only if $C \equiv \left(\begin{smallmatrix}0&0\\0&0\end{smallmatrix}\right)$. But $C + C^T = \Delta a_2 a_4 P = \left(\begin{smallmatrix}0&0\\0&0\end{smallmatrix}\right)$, and we take C to be upper triangular. Hence $C = \left(\begin{smallmatrix}0&0\\0&0\end{smallmatrix}\right) = a_4^2\overline{B}A'_{\infty^\tau}\overline{B}^T$, and in all cases we have

$$C \equiv a_4^2\overline{B}A'_{(a_2/a_4)^\tau}\overline{B}^T. \tag{2.34}$$

Now consider $t = \infty$, so $\bar{t} = \infty(a_1/a_3)^\tau$. Arguing exactly as in the case $t = 0$, we find that

$$E \equiv a_3^2\overline{B}A'_{(a_1/a_3)^\tau}\overline{B}^T. \tag{2.35}$$

This completes a proof of the following theorem (use $\overline{B}P\overline{B}^T = \Delta P$).

Theorem 2.3.4. *Let $\sigma \in Aut(F)$, $\overline{A} = \left(\begin{smallmatrix}a_4&a_2\\a_3&a_1\end{smallmatrix}\right) \in GL(2,q)$, $\overline{B} \in GL(2,q)$, $\lambda = \det(\overline{A})$, $\Delta = \det(\overline{B})$. Suppose that $\theta = \theta(\sigma, \overline{A}\otimes\overline{B}, C, E)$ as in Eq. (2.29) is an automorphism of $G^{\otimes}$ mapping the 4-gonal family $\mathcal{J}(\mathcal{C})$ to the 4-gonal family $\mathcal{J}(\mathcal{C}')$, where both $\mathcal{C}$ and $\mathcal{C}'$ are τ^{-1} normalized. Here we suppose τ is a multiplicative isomorphism of $\tilde{F}$, so that $0^\tau = 0$, $1^\tau = 1$, $\infty^\tau = \infty$, and τ commutes with elements of $Aut(F)$. Then the following hold:*

(i) $$\begin{pmatrix} C & D \\ 0 & E \end{pmatrix} = (I\otimes\overline{B})\begin{pmatrix} a_4^2A'_{(a_2/a_4)^\tau} & a_2a_3P \\ 0 & a_3^2A'_{(a_1/a_3)^\tau}\end{pmatrix}(I\otimes\overline{B})^T.$$

(ii) $$\begin{aligned}\theta(\sigma,\overline{A}\otimes\overline{B},C,E) &= \theta(\sigma,\overline{A}\otimes\overline{B}) : ((\alpha,\beta),c) \mapsto \\ &((\alpha^\sigma,\beta^\sigma)(\overline{A}\otimes\overline{B}), \lambda\Delta c^\sigma + a_4^2\alpha^\sigma(\overline{B}A'_{(a_2/a_4)^\tau}\overline{B}^T)(\alpha^\sigma)^T \\ &\quad + \Delta a_2a_3(\alpha\circ\beta)^\sigma + a_3^2\beta^\sigma(\overline{B}A'_{(a_1/a_3)^\tau}\overline{B}^T)(\beta^\sigma)^T).\end{aligned}$$

(iii) $\theta : A(t) \to A'(\bar{t})$, *where* $\bar{t} = \left(\frac{a_1t^\sigma/\tau a_2}{a_3t^\sigma/\tau a_4}\right)^\tau$.

Note. In determining the specific form of C and E we used only the effect of $\theta = \theta(\sigma, \overline{A}\otimes\overline{B}, C, E)$ on $A(0)$ and on $A(\infty)$. So the conditions given in Theorem 2.3.4 are necessary but in general not sufficient for θ to map $\mathcal{J}(\mathcal{C})$ to $\mathcal{J}(\mathcal{C}')$. In the next two sections we are going to establish that each automorphism of $GQ(\mathcal{C})$ that fixes the points (∞) and $(0,0,0)$ is induced by an automorphism of $G^{\otimes}$ of the form $\theta = \theta(\sigma, \overline{A}\otimes\overline{B})$ as in Eq. (2.29), where by Theorem 2.3.4 we no longer need to indicate the matrices C and E. And using our present point of view it seems feasible to attempt to find necessary and sufficient conditions on $\sigma, \overline{A}, \overline{B}$ for θ to map $\mathcal{J}(\mathcal{C})$ to $\mathcal{J}(\mathcal{C}')$.

Lemma 2.3.5. *Suppose $\mathcal{C}$ is τ^{-1}-normalized. For $(0,0) \neq (x,y) \in F^2$, $(0,0) \neq \alpha \in F^2$, $c \in F$, the only member of $\mathcal{J}(\mathcal{C})$ that could contain $((x,y)\otimes\alpha, c)$ is $A((y/x)^\tau)$. And $((x,y)\otimes\alpha,c) \in A((y/x)^\tau)$ iff $c = x^2g(\alpha,(y/x)^\tau)$.*

Proof. For $x = 0$, $((0,y)\otimes\alpha,c) = (0,y\alpha,c)$ is in $A^*(\infty)$, and is in $A(\infty)$ iff $c = 0 = 0\cdot g(\alpha,\infty)$. For $x \neq 0$, $((x,y)\otimes\alpha,c) = ((1,y/x)\otimes x\alpha,c) \in A^*((y/x)^\tau)$. And $((x,y)\otimes\alpha,c) \in A((y/x)^\tau)$ iff $c = (x\alpha)A_{(y/x)^\tau}(x\alpha)^\tau = x^2 g(\alpha,(y/x)^\tau)$. □

For $0 \neq t \in F$, consider the image under $\theta(\sigma,\overline{A}\otimes\overline{B})$ of the typical element of $A(t)$.

$$\begin{aligned}
&\theta(\sigma,\overline{A}\otimes\overline{B})\colon ((1,t^{1/\tau})\otimes\alpha, g(\alpha,t)) = (\alpha, t^{1/\tau}\alpha, g(\alpha,t)) \mapsto \\
&((1,t^{\sigma/\tau})\overline{A}\otimes\alpha^\sigma\overline{B}, \lambda\Delta g(\alpha,t)^\sigma + \alpha^\sigma C(\alpha^\sigma)^T + 0 + t^{2\sigma/\tau}\alpha^\sigma E(\alpha^\sigma)^T) \\
&= ((a_4 + t^{\sigma/\tau}a_3, a_2 + t^{\sigma/\tau}a_1)\otimes\alpha^\sigma\overline{B}, \\
&\qquad \alpha^\sigma\overline{B}(\lambda\Delta\overline{B}^{-1}A_t^\sigma\overline{B}^{-T} + a_4^2 A'_{(a_2/a_4)^\tau} + t^{2\sigma/\tau}a_3^2 A'_{(a_1/a_3)^\tau})(\alpha^\sigma\overline{B})^T).
\end{aligned}$$

This element of $G^\otimes$ must be in $A'_{\bar t}$ where $\bar t = \left(\frac{a_1t^{\sigma/\tau}+a_2}{a_3t^{\sigma/\tau}+a_4}\right)^\tau$, and by Lemma 2.3.5 this holds iff

$$\begin{aligned}
&\alpha^\sigma\overline{B}[\lambda\Delta\overline{B}^{-1}A_t\overline{B}^{-T} + a_4^2 A'_{(a_2/a_4)^\tau} + t^{2\sigma/\tau}a_3^2 A'_{(a_1/a_3)^\tau}](\alpha^\sigma\overline{B})^T \\
&\qquad\qquad = (a_4^2 + t^{2\sigma/\tau}a_3^2)\alpha^\sigma\overline{B}A'_{\left(\frac{a_1t^{\sigma/\tau}+a_2}{a_3t^{\sigma/\tau}+a_4}\right)^\tau}(\alpha^\sigma\overline{B})^T.
\end{aligned}$$

As this holds for all $\alpha \in F^2$, we have proved the following result.

Theorem 2.3.6. *Let $\sigma \in Aut(F)$, $\overline{A} = \left(\begin{smallmatrix} a_4 & a_2 \\ a_3 & a_1 \end{smallmatrix}\right) \in GL(2,q)$, $\overline{B} \in GL(2,q)$, $\lambda = \det(\overline{A})$, $\Delta = \det(\overline{B})$. Then $\theta = \theta(\sigma,\overline{A}\otimes\overline{B})$ (as in (ii) of Theorem 2.3.4) is an automorphism of $G^\otimes$ mapping the 4-gonal family $\mathcal{J}(\mathcal{C})$ to the 4-gonal family $\mathcal{J}(\mathcal{C}')$, where both $\mathcal{C}$ and $\mathcal{C}'$ are τ^{-1}-normalized, iff the following condition holds:*

$$\begin{aligned}
\lambda\Delta\overline{B}^{-1}A_t\overline{B}^{-T} \equiv\ & a_4^2 A'_{(a_2/a_4)^\tau} + t^{2\sigma/\tau}a_3^2 A'_{(a_1/a_3)^\tau} \\
& + (a_4^2 + t^{2\sigma/\tau}a_3^2)A'_{\left(\frac{a_1t^{\sigma/\tau}+a_2}{a_3t^{\sigma/\tau}+a_4}\right)^\tau}
\end{aligned} \tag{2.36}$$

for all $t \in F$, $t \neq 0$. (Note that C and E were determined in Theorem 2.3.4 so that the cases $t = 0$, $t = \infty$ are handled.)

It is an easy exercise to check that the condition in Eq. (2.36) imposes three separate conditions arising from the three possibly nonzero entries in the matrices of the q-clans, and that the condition on the entries in the $(1,2)$ position is automatically satisfied. So there are really the following two conditions:

1. $\frac{\lambda}{\Delta}(b_1^2x_t + b_1b_2t^{\tau^{-1}} + b_2^2z_t) = a_4^2x'_{\left(\frac{a_2}{a_4}\right)^\tau} + t^{\frac{2\sigma}{\tau}}a_3^2x'_{\left(\frac{a_1}{a_3}\right)^\tau} + (a_4^2 + t^{\frac{2\sigma}{\tau}}a_3^2)x'_{\left(\frac{a_1t^{\frac{\sigma}{\tau}}+a_2}{a_3t^{\frac{\sigma}{\tau}}+a_4}\right)^\tau}$

and

2. $\frac{\lambda}{\Delta}(b_3^2x_t + b_3b_4t^{\tau^{-1}} + b_4^2) = a_4^2z'_{\frac{a_2}{a_4}} + t^{\frac{2\sigma}{\tau}}a_3^2z'_{\frac{a_1}{a_3}} + (a_4^2 + t^{\frac{2\sigma}{\tau}}a_3^2)z'_{\frac{a_1t^{\frac{\sigma}{\tau}}+a_2}{a_3t^{\frac{\sigma}{\tau}}+a_4}}.$

Definition 2.3.7. *Put* $\mathcal{N} = \{\theta(id, I \otimes aI) : 0 \neq a \in F\}$*. Then* $\mathcal{N}$ *is a subgroup of* $Aut(G^{\otimes})$ *called the* q-clan kernel.

The group $\mathcal{N}$ deserves the name kernel for several reasons. For example, suppose that $\mathcal{C} = \mathcal{C}'$ in the statement of the Fundamental Theorem. Then there is a homomorphism $T : \theta \mapsto T_\theta$ from the group $\mathcal{H}$ to the subgroup of $P\Gamma L(4, q)$ leaving invariant the cone K and the flock $\mathcal{F}(\mathcal{C})$. The kernel of T is $\mathcal{N}$. Later on we shall return to the theme of justifying the name *kernel* for $\mathcal{N}$. For now, we note that $\theta(id, I \otimes aI) : (\alpha, \beta, c) \mapsto (a\alpha, a\beta, a^2c)$ is an automorphism of $G^{\otimes}$ that leaves invariant each $A(t)$ for $t \in \tilde{F}$ no matter what q-clan $\mathcal{C}$ is used to obtain $A(t)$.

In the present context Eq. (2.25) takes the following form:

$$\theta(\sigma_1, A_1 \otimes B_1) \circ \theta(\sigma_2, A_2 \otimes B_2) = \theta(\sigma_1 \circ \sigma_2, A_1^{\sigma_2} A_2 \otimes B_1^{\sigma_2} B_2). \tag{2.37}$$

Hence it is easy to check that

$$\theta(\sigma, A \otimes B)^{-1} = \theta(\sigma^{-1}, A^{-\sigma^{-1}} \otimes B^{-\sigma^{-1}}) \tag{2.38}$$

and

$$\theta(\sigma, A \otimes B) \cdot \theta(id, I \otimes aI) \cdot \theta(\sigma, A \otimes B)^{-1} = \theta(id, I \otimes a^{\sigma^{-1}} I). \tag{2.39}$$

Definition 2.3.8. $\theta(\sigma, A \otimes B)$ *is* linear *provided* $\sigma = id$.

Definition 2.3.9. $\theta(\sigma, A \otimes B)$ *is called* special *provided* $\Delta = \det(B) = 1$.

Note that $\{\theta(\sigma, I \otimes I) : \sigma \in Aut(F)\} \cong Aut(F)$.

Each $\theta(\sigma, A \otimes B)$ has the following unique decomposition:

$$\begin{array}{cccccccc} & & Aut(F) & & \text{special linear} & & \text{kernel} \\ & & \updownarrow & & \updownarrow & & \updownarrow \\ \theta(\sigma, A \otimes B) & = & \theta(\sigma, I \otimes I) & \cdot & \theta(id, A \otimes \Delta^{-1/2} B) & \cdot & \theta(id, I \otimes \Delta^{1/2} I)\,. \end{array} \tag{2.40}$$

Since F has characteristic 2,

$$\lambda B^{-1} A_t^{\sigma} B^{-T} = \lambda/\Delta (\Delta^{-1/2} B)^{-1} A_t^{\sigma} (\Delta^{-1/2} B)^{-T},$$

and $\Delta^{-1/2} B \in SL(2, q)$. And we observe that to determine all isomorphisms $\theta : GQ(\mathcal{C}) \to GQ(\mathcal{C}')$ mapping $(\infty), A(\infty)$, $(0,0,0)$ to (∞), $A'(\infty)$, $(0,0,0)$, respectively, we need to determine only the *special* isomorphisms $\theta(\sigma, A \otimes B)$ (i.e., $\Delta = \det(B) = 1$). Hence we have the following slightly refined version of that part of the Fundamental Theorem dealing with GQ derived from τ^{-1}-normalized q-clans.

Theorem 2.3.10 (The Fundamental Theorem refined). *Let* $\mathcal{C}$ *and* $\mathcal{C}'$ *be two* τ^{-1}*-normalized* q*-clans. Then, modulo the* q*-clan kernel* $\mathcal{N}$*, each isomorphism* $\theta : GQ(\mathcal{C}) \to GQ(\mathcal{C}')$ *mapping* (∞), $A(\infty)$, $(0,0,0)$ *to* (∞), $A'(\infty)$, $(0,0,0)$*, respectively, is given by an automorphism* $\theta : G^{\otimes} \to G^{\otimes}$ *mapping* $\mathcal{J}(\mathcal{C})$ *to* $\mathcal{J}(\mathcal{C}')$ *and* $A(\infty)$ *to* $A'(\infty)$*. There is a unique such* θ *for each* 4*-tuple* $(\lambda, B, \sigma, \pi) \in F^* \times SL(2,q) \times Aut(F) \times Sym(F)$ *for which*

(i) $A'_{\bar{t}} \equiv \lambda B^{-1} A^{\sigma}_{t} B^{-T} + A'_{\bar{0}}$ *for all* $t \in F$;

(ii) $\theta : A(t) \to A'(\bar{t})$, *where* $\pi : t \mapsto \bar{t}$ *satisfies* $\bar{t} = (\lambda t^{\sigma/\tau} + \bar{0}^{1/\tau})^{\tau}$.

The associated automorphism $\theta : G^{\otimes} \to G^{\otimes}$ *is given by*

(iii) $\theta = \theta(\sigma, \begin{pmatrix} 1 & \bar{0}^{1/\tau} \\ 0 & \lambda \end{pmatrix} \otimes B)$:

$$((\alpha, \beta), c) \mapsto ((\alpha^{\sigma}, \beta^{\sigma}) \left(\begin{pmatrix} 1 & \bar{0}^{1/\tau} \\ 0 & \lambda \end{pmatrix} \otimes B \right), \lambda c^{\sigma} + g'(\alpha^{\sigma} B, \bar{0})).$$

Here $g'(\alpha^{\sigma} B, \bar{0}) = \alpha^{\sigma} B A'_{\bar{0}} B^{T} (\alpha^{\sigma})^{T}$.

In almost all the applications of the F. T. in this book, the q-clans involved are 1/2-normalized, i.e., $\tau = 2$. *We note that in this case* (ii) *takes the following form:*

(ii)$'$ $\bar{t} = \lambda^2 t^{\sigma} + \bar{0}$ (*when* $\tau = 2$).

Note. When $\tau = 2$ and θ is linear, so $\bar{t} = \lambda^2 t + \bar{0}$, if $t = \bar{t}$ with $t \in F$ and λ not equal to 1, then $t = \frac{\bar{0}}{1+\lambda^2}$, so t is unique! Hence if $t = \bar{t}$ for two values of $t \in F$ it must be that $\bar{t} = t + \bar{0}$, which fixes no element of F or all elements of F.

So suppose that the q-clan $\mathcal{C}$ is 1/2-normalized. If we have an appropriate 4-tuple $(\lambda, B, \sigma, \pi)$ for which

$$A_{\bar{t}} \equiv \lambda B^{-1} A^{\sigma}_{t} B^{-T} + A_{\bar{0}} \text{ for all } t \in F, \tag{2.41}$$

then the associated automorphism $\theta : G^{\otimes} \to G^{\otimes}$ (which induces a collineation of $GQ(\mathcal{C})$) is given by

$$\theta = \theta(\sigma, \begin{pmatrix} 1 & \bar{0}^{\frac{1}{2}} \\ 0 & \lambda \end{pmatrix} \otimes B) : ((\alpha, \beta), c) \mapsto$$
$$((\alpha^{\sigma}, \beta^{\sigma}) \begin{pmatrix} 1 & \bar{0}^{\frac{1}{2}} \\ 0 & \lambda \end{pmatrix} \otimes B, \lambda c^{\sigma} + g(\alpha^{\sigma} B, \bar{0})). \tag{2.42}$$

Here $g(\alpha^{\sigma} B, \bar{0}) = \alpha^{\sigma} B A_{\bar{0}} B^{T} (\alpha^{\sigma})^{T}$.

Since

$$(1, t^{\frac{1}{2}})^{\sigma} \begin{pmatrix} 1 & \bar{0}^{\frac{1}{2}} \\ 0 & \lambda \end{pmatrix} = (1, (\lambda^2 t^{\sigma} + \bar{0})^{\frac{1}{2}}),$$

we see that

$$\theta : [A(t)] \mapsto [A(\bar{t})], \text{ where } \bar{t} = \lambda^2 t^{\sigma} + \bar{0} \text{ for all } t \in F. \tag{2.43}$$

Similarly, if $B = \left(\begin{smallmatrix} b4 & b2 \\ b3 & b1 \end{smallmatrix}\right)$, then

$$(1, s^{\frac{1}{2}})^{\sigma} B \equiv (1, \left(\frac{b_1^2 s^{\sigma} + b_2^2}{b_3^2 s^{\sigma} + b_4^2} \right)^{\frac{1}{2}}).$$

Hence

$$\mathcal{O}_s \mapsto \mathcal{O}'_{\bar{s}} \text{ where } \bar{s} = \frac{b_1^2 s^\sigma + b_2^2}{b_3^2 s^\sigma + b_4^2}. \tag{2.44}$$

Since $\theta(\sigma, A \otimes B) : (\gamma \otimes \alpha, c) \mapsto (\gamma^\sigma A \otimes \alpha^\sigma B, c')$, where we are not ready to compute c', clearly

$$\theta(\sigma, A \otimes B) : \mathcal{L}_\gamma \to \mathcal{L}_{\gamma^\sigma A} \text{ and } \theta(\sigma, A \otimes B) : \mathcal{R}_\alpha \to \mathcal{R}_{\alpha^\sigma B}. \tag{2.45}$$

For efficiency in calculating later, we also note that if $A = \left(\begin{smallmatrix} a4 & a2 \\ a3 & a1 \end{smallmatrix}\right)$ and B is as above, then

$$\theta(\sigma, A \otimes B) : A(t) \to A(\bar{t}) \text{ where } (1, t^\sigma)A^{(2)} \equiv (1, \bar{t}), \tag{2.46}$$

and

$$\theta(\sigma, A \otimes B) : \mathcal{O}_s \to \mathcal{O}_{\bar{s}} \text{ where } (1, s^\sigma)B^{(2)} \equiv (1, \bar{s}). \tag{2.47}$$

2.4 Extension to 1/2-Normalized q-Clans

Let $\mathcal{C} = \{A_t = \begin{pmatrix} xt & t^{1/2} \\ 0 & zt \end{pmatrix} : t \in F\}$ be a 1/2-normalized q-clan (so $A_0 = \left(\begin{smallmatrix} 0 & 0 \\ 0 & 0 \end{smallmatrix}\right)$). The Subiaco and Adelaide q-clans, to be defined later and which are probably our main interest, are always given in 1/2-normalized form. But the other known examples are also easily given in this form.

Any automorphism θ of $G^\otimes$ replaces the 4-gonal family $\mathcal{J} = \mathcal{J}(\mathcal{C})$ with some 4-gonal family $\mathcal{J}^\theta$. But we are especially interested in certain types of automorphisms of $G^\otimes$ that produce 4-gonal families easily seen to have associated q-clans.

For each $s \in F$, define $\tau_s : G^\otimes \to G^\otimes$ (*shift by s*) by

$$\begin{aligned} \tau_s = \theta\left(id, \begin{pmatrix} 1 & y_s \\ 0 & 1 \end{pmatrix} \otimes I\right) &: ((\alpha, \beta), c) \mapsto \\ &((\alpha, \beta)\left[\begin{pmatrix} 1 & y_s \\ 0 & 1 \end{pmatrix} \otimes I\right], c + \alpha A_s \alpha^T) = (\alpha, \beta + y_s\alpha, c + g(\alpha, s)). \end{aligned} \tag{2.48}$$

In particular,

$$\begin{aligned} \tau_s : (\gamma_{yt} \otimes \alpha, g(\alpha, t)) &\mapsto ((\gamma_{yt} \otimes \alpha)\left[\begin{pmatrix} 1 & y_s \\ 0 & 1 \end{pmatrix} \otimes I\right], g(\alpha, t) + g(\alpha, s)) \\ &= (\gamma_{yt+ys} \otimes \alpha, \alpha(A_t + A_s)\alpha^T)) = (\gamma_{yt+s} \otimes \alpha, g^{\tau s}(\alpha, t + s)). \end{aligned}$$

Keep in mind that $y_\infty = 0$, $\gamma_{y\infty} = (0, 1)$, and $s + \infty = \infty$ for $s \in F$. So the new q-clan $\mathcal{C}^{\tau s} = \{A_t^{\tau s} : t \in F\}$ satisfies

$$A_t^{\tau s} = A_{t+s}^{\tau s} = A_t + A_s, \text{ i.e., } A_x^{\tau s} = A_{x+s} + A_s, \quad x \in F. \tag{2.49}$$

So clearly $\mathcal{C} \sim \mathcal{C}^{\tau_s}$ by an equivalence in which $\lambda = 1$, $B = I$, $\sigma = id$, and $\pi : t \mapsto \bar{t} = t + s$. Then $g^{\tau_s}(\alpha^\sigma B, \bar{0}) = g_{\bar{0}}^{\tau_s}(\alpha) = \alpha A_{\bar{0}}^{\tau_s}\alpha^T = \alpha(A_{0+s}^{\tau_s})\alpha^T = \alpha(A_0 + A_s)\alpha^T = \alpha A_s \alpha^T = g(\alpha, s)$. Use this to compare Eq. (2.48) with the F.T. Also note that $y_t = t^{1/2}$ implies that $y_t^{\tau_s} = y_{t+s} + y_s = t^{1/2}$, so that $\mathcal{C}^{\tau_s}$ is also 1/2-normalized.

For $0 \neq a \in F$, define $\sigma_a : G^\otimes \to G^\otimes$ (*scale by a*) by

$$\sigma_a = \theta(id, \begin{pmatrix} 1 & 0 \\ 0 & a^{1/2} \end{pmatrix} \otimes I) : ((\alpha, \beta), c) \mapsto$$

$$((\alpha, \beta)\left[\begin{pmatrix} 1 & 0 \\ 0 & a^{1/2} \end{pmatrix} \otimes I\right], a^{1/2}c) = (\alpha, a^{1/2}\beta, a^{1/2}c). \quad (2.50)$$

Then $\sigma_a : (\gamma_{y_t} \otimes \alpha, g(\alpha, t)) \mapsto ((\gamma_{y_t} \otimes \alpha)\left[\left(\begin{smallmatrix} 1 & 0 \\ 0 & a^{1/2} \end{smallmatrix}\right) \otimes I\right]$,

$$a^{1/2}g(\alpha, t)) = (\gamma_{y_t}\left(\begin{smallmatrix} 1 & 0 \\ 0 & a^{1/2} \end{smallmatrix}\right) \otimes \alpha, \alpha \begin{pmatrix} a^{1/2}x_t & (at)^{1/2} \\ 0 & a^{1/2}z_t \end{pmatrix} \alpha^T) = (\gamma_{y_{at}} \otimes \alpha, g^{\sigma_a}(\alpha, at)).$$

Here A_t is replaced with $A_{at}^{\sigma_a} = \begin{pmatrix} a^{1/2}x_t & (at)^{1/2} \\ 0 & a^{1/2}z_t \end{pmatrix}$, so $\mathcal{C}^{\sigma_a}$ is 1/2-normalized.

Define $\varphi : G^\otimes \to G^\otimes$ (the *flip*) by

$$\varphi := \theta(id, P \otimes I) : ((\alpha, \beta), c) \mapsto ((\beta, \alpha), c + \alpha \circ \beta). \quad (2.51)$$

Here $\varphi : A(\infty) \leftrightarrow A(0)$, and for $0 \neq t \in F$,

$$\varphi : ((\gamma_{y_t} \otimes \alpha), g(\alpha, t)) \mapsto$$
$$((\gamma_{y_t} \otimes \alpha)(P \otimes I), \alpha A_t \alpha^T) = ((t^{1/2}, 1) \otimes \alpha, \alpha A_t \alpha^T)$$
$$= ((1, t^{-1/2}) \otimes t^{1/2}\alpha, (t^{1/2}\alpha)(t^{-1}A_t)(t^{1/2}\alpha)^T) = (\gamma_{y_{t^{-1}}} \otimes t^{1/2}\alpha, g^\varphi(t^{1/2}\alpha, t^{-1})).$$

As $A_t = \begin{pmatrix} x_t & t^{1/2} \\ 0 & z_t \end{pmatrix}$ is replaced with $A_{t^{-1}}^\varphi t^{-1} A_t = \begin{pmatrix} x_t/t & t^{-1/2} \\ 0 & z_t/t \end{pmatrix}$, so

$$\mathcal{C}^\varphi = \{A_0^\varphi = \left(\begin{smallmatrix} 0 & 0 \\ 0 & 0 \end{smallmatrix}\right)\} \cup \{A_{t^{-1}}^\varphi = t^{-1}A_t : 0 \neq t \in F\}$$

is 1/2-normalized.

Of the three types of automorphisms of $G^\otimes$ just considered, flipping is the only one that moves $A(\infty)$. It is clear that flipping replaces a q-clan $\mathcal{C}$ with a new q-clan $\mathcal{C}^\varphi$, but it is not clear in general whether or not $\mathcal{C} \sim \mathcal{C}^\varphi$. We shall return to this question later.

Shifting, flipping and scaling provide recoordinatizations of a given $GQ(\mathcal{C})$. As permutations of the indices of the lines through (∞), these recoordinatizations have the following descriptions as linear fractional maps on $\tilde{F}$: $\tau_s : t \mapsto t + s$; $\sigma_a : t \mapsto at$; $\varphi : t \mapsto t^{-1}$. First shifting, then flipping (or not), then shifting and scaling provide all the Möbius transformations. Hence we recognize $PGL(2, q)$ acting on

$\tilde{F} \cong PG(1,q)$. And $PGL(2,q)$ is sharply triply transitive on $PG(1,q)$. Suppose θ_1 and θ_2 are two different sequences of shifts, flips and scales that effect the same permutation on $\tilde{F}$ and replace $GQ(\mathcal{C})$ with $GQ(\mathcal{C}^{\theta_1})$ and $GQ(\mathcal{C}^{\theta_2})$, respectively. It would be nice to know that $\mathcal{C}^{\theta_1} \sim \mathcal{C}^{\theta_2}$. That this is so follows immediately as a corollary of the first theorem of the next section.

2.5 A Characterization of the q-Clan Kernel

Theorem 2.5.1. *Let $\theta : G^{\otimes} \to G^{\otimes}$ be an automorphism of $G^{\otimes}$ obtained as a finite sequence of shifts, flips and scales. Moreover, suppose $\mathcal{C}$ and $\mathcal{C}'$ are two 1/2-normalized q-clans for which $\theta : \mathcal{J}(\mathcal{C}) \to \mathcal{J}(\mathcal{C}')$ in such a way that θ effects the identity permutation on $\tilde{F}$. Then θ must have the form $\theta : (\alpha, \beta, c) \mapsto (a\alpha, a\beta, a^2 c)$ for some nonzero $a \in F$. Hence $\mathcal{J}(\mathcal{C}) \equiv \mathcal{J}(\mathcal{C}')$, and θ belongs to the q-clan kernel $\mathcal{N}$.*

Proof. Any word in shifts, scales and flips clearly has the form $\theta(id, A \otimes I)$, $A = \left(\begin{smallmatrix} a_4 & a_2 \\ a_3 & a_1 \end{smallmatrix}\right)$, with $\theta : A(t) \to A'(\bar{t})$, where $\bar{t} = \frac{a_1^2 t + a_2^2}{a_3^2 t + a_4^2}$. And $\bar{t} = t$ for all $t \in \tilde{F}$ iff $a_2 = a_3 = 0$ and $a_1 = a_4 \neq 0$, i.e., $A = aI$ for some nonzero $a \in F$. Hence $\theta(id, A \otimes I) = \theta(id, aI \otimes I) = \theta(id, I \otimes aI) \in \mathcal{N}$. □

Following Theorem 2.3.6 we made the claim that the following theorem is true.

Theorem 2.5.2. *Suppose that in the F.T. (Theorem 2.2.1) the two q-clans $\mathcal{C}$, $\mathcal{C}'$ are 1/2-normalized and identical, i.e., $A'_t \equiv A_t$ for all $t \in F$. Then the map $T : \theta \mapsto T_\theta$ is a homomorphism from the group $\mathcal{H}$ (of collineations of $GQ(\mathcal{C})$ fixing (∞), $[A(\infty)]$ and $(0,0,0)$) onto the subgroup of $P\Gamma L(4,q)$ leaving invariant the cone K and the flock $\mathcal{F}(\mathcal{C})$. The kernel of T is the q-clan kernel $\mathcal{N}$.*

Proof. The original claim was made for q-clans in general form, not just those that are 1/2-normalized. So we write out the proof for the general form.

Each $\theta \in \mathcal{H}$ is an automorphism of $G^{\otimes}$ of the form $\theta = \theta(\lambda, B, \sigma, \pi)$. We have been using the notation $\pi : t \mapsto \bar{t}$, so that the image of 0 under π is $\bar{0}$. Throughout this proof, since we will have more than one permutation π involved, we need to write 0^{π} in place of $\bar{0}$, etc. For example, to say that $\theta_1 = \theta(\lambda_1, B_1, \sigma_1, \pi_1)$ is in $\mathcal{H}$ means that

$$A_{t^{\pi_1}} \equiv \lambda_1 B_1^{-1} A_t^{\sigma_1} B_1^{-T} + A_{0^{\pi_1}} \text{ for all } t \in F. \tag{2.52}$$

And the action of θ_1 is given by the following, where $\Delta_1 = \det(B_1)$ and $\overline{A}_1 = \begin{pmatrix} 1 & y_{0^{\pi_1}} \\ 0 & \lambda_1/\Delta_1 \end{pmatrix}$:

$$\theta_1 = \theta(\sigma_1, \overline{A}_1 \otimes B_1) : \; ((\alpha, \beta), c) \mapsto ((\alpha^{\sigma}, \beta^{\sigma})(\overline{A}_1 \otimes B_1), \lambda_1 c^{\sigma_1} + g'(\alpha^{\sigma} B, 0^{\pi_1})). \tag{2.53}$$

For such a θ_1 we can define the action of T_{θ_1} on the planes of the flock $\mathcal{C}$ as follows, where $B_1^{-1} = \left(\begin{smallmatrix} a1 & b1 \\ c1 & d1 \end{smallmatrix}\right)$:

$$T_{\theta_1}\colon \begin{bmatrix} x_t \\ y_t \\ z_t \\ 1 \end{bmatrix} \mapsto \begin{bmatrix} x_{t^\pi} \\ y_{t^\pi} \\ z_{t^\pi} \\ 1 \end{bmatrix} = \begin{pmatrix} \lambda_1 a_1^2 & \lambda_1 a_1 b_1 & \lambda_1 b_1^2 & x_{0^{\pi_1}} \\ 0 & \lambda_1(a_1 d_1 + b_1 c_1) & 0 & y_{0^{\pi_1}} \\ \lambda_1 c_1^2 & \lambda_1 c_1 d_1 & \lambda_1 d_1^2 & z_{0^{\pi_1}} \\ 0 & 0 & 0 & 1 \end{pmatrix} \begin{bmatrix} x_t^{\sigma 1} \\ y_t^{\sigma 1} \\ z_t^{\sigma 1} \\ 1 \end{bmatrix}. \tag{2.54}$$

Keep in mind that Eq. (2.52) is really just an efficient encoding of the description of T_{θ_1} given in Eq. (2.54). Suppose we have a second member θ_2 of $\mathcal{H}$ with λ_2, σ_2, etc. We want to show that $T_{\theta_1} \circ T_{\theta_2} = T_{\theta_1 \circ \theta_2}$. To do this we use Eq. (2.52) instead of Eq. (2.54) to compute $T_{\theta_1} \circ T_{\theta_2}$. And we use either Eq. (2.37) or the definition of θ_i as in Eq. (2.53) to compute $T_{\theta_1 \circ \theta_2}$. In the former case we have

$$\begin{aligned} A_{(t^{\pi_1})^{\pi_2}} &\equiv \lambda_2 B_2^{-1} [A_{t^{\pi_1}}]^{\sigma 2} B_2^{-T} + A_{0^{\pi_2}} \\ &\equiv \lambda_2 B_2^{-1} [\lambda_1 B_1^{-1} A_t^{\sigma 1} B_1^{-T} + A_{0^{\pi_1}}]^{\sigma 2} B_2^{-T} + A_{0^{\pi_2}} \\ &\equiv \lambda_1^{\sigma 2} \lambda_2 B_2^{-1} B_1^{-\sigma 2} A_t^{\sigma 1 \sigma 2} B_1^{-T \sigma 2} B_2^{-T} + \lambda_2 B_2^{-1} A_{0^{\pi_1}}^{\sigma 2} B_2^{-T} + A_{0^{\pi_2}} \\ &\equiv (\lambda_1^{\sigma 2} \lambda_2)(B_1^{\sigma 2} B_2)^{-1} A_t^{\sigma 1 \sigma 2} (B_1^{\sigma 2} B_2)^{-T} + A_{0^{\pi_1 \circ \pi_2}}. \end{aligned}$$

In the latter case we use $A_i = \left(\begin{smallmatrix} 1 & y_{0^{\pi_i}} \\ 0 & \lambda_i/\Delta_i \end{smallmatrix}\right)$ for $i = 1$ and $i = 2$ and then compute $\theta_1 \circ \theta_2$ using Eq. (2.37): (Warning: The notation $\overline{A}_i$ was intended to avoid confusing this matrix with a matrix of the q-clan, but it gets a bit clumsy. Here we have chosen to eliminate the overline and hope that the reader will realize that a matrix A_i with $i = 1$ or $i = 2$ is a matrix encoding part of the action of a collineation and not a matrix of the q-clan.) Recall that

$$\theta(\sigma_1, A_1 \otimes B_1) \circ \theta(\sigma_2, A_2 \otimes B_2) = \theta(\sigma_1 \circ \sigma_2, A_1^{\sigma 2} A_2 \otimes B_1^{\sigma 2} B_2). \tag{2.55}$$

By the Fundamental Theorem this latter collineation, which clearly must fix (∞), $[A(\infty)]$ and $(0,0,0)$ and is hence in $\mathcal{H}$, must be a collineation of the form $\theta(\lambda, B, \sigma, \pi)$. From the above computations we see that $\lambda = \lambda_1^{\sigma 2}\lambda_2$, $B = B_1^{\sigma 2} B_2$, $\sigma = \sigma_1 \circ \sigma_2$, and $\pi \pi_1 \circ \pi_2$. It is now easy to see that $T : \theta \mapsto T_\theta$ is a homomorphism. It is also easy to check that the kernel is $\mathcal{N}$. The fact that it is onto the subgroup stabilizing the flock is really a corollary of the Fundamental Theorem.

It is amusing first to use Observation 1.2.2 and Eq. (2.52) to check that

$$y_{0^{\pi_1 \circ \pi_2}} = (y_{0^{\pi_1}})^{\sigma 2} \lambda_2 / \Delta_2 + y_{0^{\pi_2}}, \tag{2.56}$$

and then to derive the same formula by computing the upper right hand entry of the matrix $A_1^{\sigma 2} A_2$. □

The preceding theorem provides some justification for calling $\mathcal{N}$ the q-clan kernel. But for *nonlinear* flocks, i.e., for every non-classical q-clan $\mathcal{C}$, the Theorem

2.5.4 shows that $\mathcal{N}$ plays a role similar to the one played by the kernel of a translation GQ (cf. 8.5 of [PT84]), so there is even more reason to call $\mathcal{N}$ the q-clan kernel. In order to prove Theorem 2.5.4, we need to recall the following result by J. A. Thas [Th87]:

Theorem 2.5.3. *Let $q = 2^e$. If all the planes of a flock share a common point, then the flock is linear.*

Proof. Embed the cone K of the flock in the hyperbolic quadric $Q^+(5,q)$ such that $Q^+(5,q) \cap H = K$, where $H \simeq PG(3,q)$. Let $\perp$ be the polarity of $PG(5,q)$ induced by $Q^+(5,q)$. Let $\pi_t^\perp$ be the polar plane of π_t with respect to $Q^+(5,q)$ and $\pi_t^\perp \cap Q^+(5,q)$ be the conic $C_t^\perp$. Since $P \in \pi_t \subseteq H$ for all $t \in F^*$, then $H^\perp \subseteq \pi_t^\perp \subseteq P^\perp$, i.e., each polar plane $\pi_t^\perp$ contains the tangent line $L = H^\perp$ to $Q^+(5,q)$ and is contained in $P^\perp = PG(4,q)$. Intersect now with $Q^+(5,q)$, then no two points in $\mathcal{O}^* = \cup_{t=1}^q C^\perp{}_t$ are collinear in $Q^+(5,q)$ because π_t and π_r, $t \neq r$, meet in an exterior line to $Q^+(5,q)$ and hence $\pi_t^\perp$ and $\pi_r^\perp$ belong to a $PG(3,q)$ intersecting $Q^+(5,q)$ in an elliptic quadric. So $\mathcal{O}^*$ is an ovoid of $Q^+(5,q)$ given by the union of q conics through the vertex of the cone, which is also an ovoid of $Q(4,q) = P^\perp \cap Q^+(5,q)$. Note that P is the nucleus of $Q(4,q)$ (see [Hi98]). As in [Th87], for any two distinct points z_1, $z_2 \in \mathcal{O}^*$, the plane spanned by P, z_1, z_2 meets $\mathcal{O}^*$ only in the points z_1 and z_2. Project now $\mathcal{O}^*$ from P into a $PG(3,q) \subseteq PG(4,q)$ not containing P. No three projected points are collinear and so they form an ovoid of $PG(3,q)$ given by the union of the q conics, each obtained as a projection of the conic $C_t^\perp$.Relying on a result by Brown [Br00], the projected ovoid is an elliptic quadric, hence also $\mathcal{O}^*$ is an elliptic quadric, i.e., $\mathcal{O}^* = Q^-(3,q) = S \cap Q^+(5,q)$, where $S \simeq PG(3,q)$. Hence $\pi_t^\perp \subset S$, for $t = 1, ..., q$. Applying the polarity $\perp$, $S^\perp \subset \pi_t$, i.e., all the planes π_t of the flock contain the common line $S^\perp$, thus the flock is linear. □

We are now ready to prove the following:

Theorem 2.5.4. *When $\mathcal{C}$ is a nonlinear $1/2$-normalized q-clan, the kernel $\mathcal{N}$ is the group of collineations of $GQ(\mathcal{C})$ fixing (∞) and $(0,0,0)$ linewise.*

Proof. Clearly each element of $\mathcal{N}$ fixes (∞) and $(0,0,0)$ linewise. So let θ be any collineation of $GQ(\mathcal{C})$ that does so. Then in the notation of the F. T. (Theorem 2.2.1), $\sigma = id$, $\bar{0} = 0$, $\pi = id$, $\lambda = 1$, and $A_t \equiv B^{-1}A_tB^{-T}$ for all $t \in F$, for some $B \in SL(2,q)$ (i.e., we are reducing θ modulo $\mathcal{N}$). Since the lines through the point (∞) correspond to the planes of the flock $\mathcal{F}(\mathcal{C})$ and θ fixes (∞) linewise, θ has to leave invariant also each plane of the flock. Among them, also the plane π_0 must be fixed, hence we are allowed to deal with the 3×3-matrix $\begin{pmatrix} a^2 & ab & b^2 \\ 0 & 1 & 0 \\ c^2 & cd & d^2 \end{pmatrix}$. Suppose $B^{-1} = \left(\begin{smallmatrix} a & b \\ c & d \end{smallmatrix}\right)$ with $\Delta = ad + bc = 1$, so imposing the condition that each plane of the flock be fixed, we have

$$\begin{pmatrix} a^2 & ab & b^2 \\ 0 & 1 & 0 \\ c^2 & cd & d^2 \end{pmatrix} \begin{pmatrix} x_t \\ y_t \\ z_t \end{pmatrix} = \begin{pmatrix} x_t \\ y_t \\ z_t \end{pmatrix}$$

for all $t \in F$.

We claim that this implies that $I = \begin{pmatrix} a^2 & ab & b^2 \\ 0 & 1 & 0 \\ c^2 & cd & d^2 \end{pmatrix}$. For if not, then there are elements $u, v, w \in F$, not all zero, for which the point $P = (u, v, w, 0) \in PG(3, q)$ lies on each plane $\pi_t = [x_t, y_t, z_t, 1]^T$ of the flock $\mathcal{F}(\mathcal{C})$.
Relying now on Theorem 2.5.3, the flock has to be linear and hence there is a contradiction. It follows that $B^{-1} = I$, so $\theta \equiv id$ (mod $\mathcal{N}$). □

If $\mathcal{C}$ is nonlinear and $\theta(\sigma, A \otimes B)$ fixes all lines through (∞), then $\sigma = id$ and $A \otimes B = I \otimes aI = aI \otimes I$. If $B \in SL(2, q)$, then $A = aI$ for $0 \neq a \in F$.

2.6 Very Important Concept

Very Important Concept. To each line through (∞) in $GQ(\mathcal{C})$ we are about to assign a class of projectively equivalent flocks (i.e., an equivalence class of q-clans). For each $s \in F$, let $i_s = \tau_s \circ \varphi$, a shift by s followed by the flip φ, i.e., a *shift-flip*. And put $i_\infty = id : G^\otimes \to G^\otimes$. Start with a 1/2-normalized q-clan $\mathcal{C}$. For each $s \in \tilde{F}$, applying i_s to $G^\otimes$ yields a 1/2-normalized q-clan $\mathcal{C}^{i_s}$. We *assign to the line* $[A(s)]$ *the class of flocks projectively equivalent to* $\mathcal{F}(\mathcal{C}^{i_s})$. Then the following theorem is basic to the theory.

Theorem 2.6.1. *Let $\mathcal{C}$ be a 1/2-normalized q-clan. Then there is an automorphism θ of $GQ(\mathcal{C})$ mapping $[A(s)]$ to $[A(t)]$, $s, t \in \tilde{F}$, iff the flocks $\mathcal{F}(\mathcal{C}^{i_s})$ and $\mathcal{F}(\mathcal{C}^{i_t})$ are projectively equivalent.*

Proof. It is an interesting but easy exercise to prove that without loss of generality we need consider only collineations θ that fix (∞) and $(0, 0, 0)$. However, for our purposes here it suffices to restate the theorem so that it refers only to such θ. Then the F. T. applied to Fig. 2.6 completes the proof. □

2.7 The q-clan $\mathcal{C}^{i_s}$, s ∈ F

We continue to suppose that $\mathcal{C}$ is $\frac{1}{2}$-normalized, so always $y_t = t^{1/2}$, $t \in F$. The image q-clan $\mathcal{C}^{i_s}$ under i_s is also naturally $\frac{1}{2}$-normalized, since each of $\mathcal{C}^{\tau_s}$ and $\mathcal{C}^{\varphi}$ are. But we prefer to consider i_s directly, rather than as a composition $\tau_s \circ \varphi$. First use Eqs. (2.48) and (2.51) to compute that for $s \in F$,

$$i_s = \theta(id, \begin{pmatrix} y_s & 1 \\ 1 & 0 \end{pmatrix} \otimes I) : ((\alpha, \beta), c) \mapsto (y_s\alpha + \beta, \alpha, c + g(\alpha, s) + \alpha \circ \beta). \quad (2.57)$$

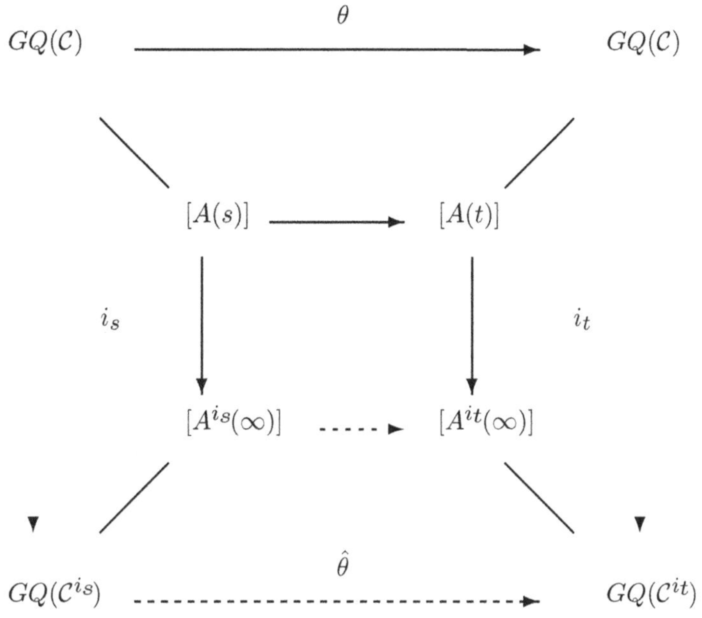

$$\hat{\theta} \;=\; i_s^{-1} \circ \theta \circ i_t \;\text{ or }\; \theta \;=\; i_s \circ \hat{\theta} \circ i_t^{-1}$$

Figure 2.6: Generalized Version of F. T.

For $t \in F$, consider the image under i_s, $s \in F$, of the typical element of $A(t)$.

$$\begin{aligned} i_s : (\gamma_{yt} \otimes \alpha, g(\alpha, t)) = (\alpha, t^{1/2}\alpha, g(\alpha, t)) \mapsto \\ ((s^{1/2} + t^{1/2})\alpha, \alpha, g(\alpha, t) + g(\alpha, s)) \\ = ((1, (s+t)^{-1/2}) \otimes (y_s + y_t)\alpha, g(\alpha, t) + g(\alpha, s)) \\ = (\gamma_{(s+t)^{-1/2}} \otimes \beta, (y_s + y_t)^{-2}(g(\beta, t) + g(\beta, s)), \end{aligned}$$

where $\beta = (y_s + y_t)\alpha$.

This element of $G^{\otimes}$ must be in $A^{is}((s+t)^{-1})$, since we want $\mathcal{C}^{is}$ to be $\frac{1}{2}$-normalized. Hence the matrix $A^{is}_{(s+t)^{-1}}$ must equal $(s+t)^{-1}(A_t + A_s)$. Put $x = (s+t)^{-1}$, so $t = s + x^{-1}$. Then

$$A^{is}_x = x(A_{s+x^{-1}} + A_s), \text{ for all } x, s \in F. \tag{2.58}$$

At this point we can compare Eq. (2.58) with Corollary 1.4.4 to see that the $q+1$ possibly different q-clans produced back in Corollary 1.4.4 are exactly those attached to the $q+1$ lines through (∞). Hence we now know that two of them are equivalent if and only if the corresponding lines through (∞) are in the same orbit.

By the Fundamental Theorem 2.3.10 , the most general special automorphism of $G^{\otimes}$ mapping $\mathcal{J}(\mathcal{C})$ to $\mathcal{J}(\mathcal{C}^{is})$ and $A(\infty)$ to $A^{is}(\infty)$ is

$$\bar{\theta} = \theta(\sigma, \begin{pmatrix} 1 & \bar{0}^{\frac{1}{2}} \\ 0 & \lambda \end{pmatrix} \otimes B),$$

where $A^{is}_{\bar{t}} + A^{is}_{\bar{0}} \equiv \lambda B^{-1} A^{\sigma}_{t} B^{-T}$ for all $t \in F$, with $0 \neq \lambda \in F$, $\sigma \in Aut(F)$, $B \in SL(2,q)$, and $\pi : t \mapsto \bar{t} = \lambda^2 t^{\sigma} + \bar{0}$. And using Eq. (2.58) we find

$$\begin{aligned} A^{is}_{\bar{t}} &+ A^{is}_{\bar{0}} A^{is}_{(\lambda^2 t^{\sigma}+\bar{0})} + A^{is}_{\bar{0}} \\ &= (\lambda^2 t^{\sigma} + \bar{0})(A_{(\frac{s\lambda^2 t^{\sigma} + s\bar{0} + 1}{\lambda^2 t^{\sigma}+\bar{0}})} + A_s) + \bar{0}(A_{s+\bar{0}^{-1}} + A_s) \\ &= (\lambda^2 t^{\sigma} + \bar{0}) A_{(\frac{s\lambda^2 t^{\sigma} + s\bar{0} + 1}{\lambda^2 t^{\sigma}+\bar{0}})} + \lambda^2 t^{\sigma} A_s + \bar{0} A_{s+\bar{0}^{-1}} \end{aligned} \tag{2.59}$$

for all $t \in F$.

In Fig. 2.6 replace s with ∞ and t with s to see that the most general automorphism θ of $GQ(\mathcal{C})$ mapping $[A(\infty)]$ to $[A(s)]$ is

$$\begin{aligned} \theta = \bar{\theta} \circ i_s^{-1} &= \theta(\sigma, \begin{pmatrix} 1 & \bar{0}^{\frac{1}{2}} \\ 0 & \lambda \end{pmatrix} \otimes B) \circ \theta(id, \begin{pmatrix} 0 & 1 \\ 1 & s^{\frac{1}{2}} \end{pmatrix} \otimes I) \\ &= \theta(\sigma, \begin{pmatrix} \bar{0}^{\frac{1}{2}} & 1 + (\bar{0}s)^{\frac{1}{2}} \\ \lambda & \lambda s^{\frac{1}{2}} \end{pmatrix} \otimes B). \end{aligned} \tag{2.60}$$

We have essentially proved the following:

Theorem 2.7.1. *The most general special automorphism of $GQ(\mathcal{C})$ fixing (∞), $(0,0,0)$ and mapping $[A(\infty)]$ to $[A(s)]$, for $s \in F$, is given by the automorphism θ of $G^{\otimes}$ defined by*

(i) $\theta = \theta(\sigma, \begin{pmatrix} \bar{0}^{\frac{1}{2}} & 1 + (\bar{0}s)^{\frac{1}{2}} \\ \lambda & \lambda s^{\frac{1}{2}} \end{pmatrix} \otimes B) : A(t) \mapsto A(\bar{t})$, *where*

(ii) $\bar{t} = (s\lambda^2 t^{\sigma} + \bar{0}s + 1)/(\lambda^2 t^{\sigma} + \bar{0})$, *subject to the condition that*

(iii) $\lambda B^{-1} A^{\sigma}_{t} B^{-T} \equiv A^{is}_{\bar{t}} + A^{is}_{\bar{0}}$ *for all* $t \in F$.

In this case

(iv) $\theta(\sigma, \begin{pmatrix} \bar{0}^{\frac{1}{2}} & 1 + (\bar{0}s)^{\frac{1}{2}} \\ \lambda & \lambda s^{\frac{1}{2}} \end{pmatrix} \otimes B) : ((\alpha, \beta, c) \mapsto$

$((\alpha^{\sigma}, \beta^{\sigma})[\begin{pmatrix} \bar{0}^{\frac{1}{2}} & 1 + (\bar{0}s)^{\frac{1}{2}} \\ \lambda & \lambda s^{\frac{1}{2}} \end{pmatrix} \otimes B]$,

$\lambda c^{\sigma} + (\alpha^{\sigma}, \beta^{\sigma})(I \otimes B) \begin{pmatrix} \bar{0} A_{s+\bar{0}^{-1}} & \lambda(1+\bar{0}s)^{\frac{1}{2}} P \\ 0 & \lambda^2 A_s \end{pmatrix} (I \otimes B^T)(\alpha^{\sigma}, \beta^{\sigma})^T)$.

Moreover, if $\bar{\theta} = \theta(\sigma, \begin{pmatrix} 1 & \bar{0} \\ 0 & \lambda^2 \end{pmatrix}^{\frac{1}{2}} \otimes B)$ *is a typical special automorphism of* $G^{\otimes}$ *mapping* $\mathcal{J}(\mathcal{C}^{is})$ *to* $\mathcal{J}(\mathcal{C}^{is})$ *and leaving* $A^{is}(\infty)$ *invariant, then*

(v) $\theta = i_s \circ \bar{\theta} \circ i_s^{-1} = \theta(\sigma, \begin{pmatrix} \lambda^2 + \bar{0}s^{\sigma} & \lambda^2 s + s^{\sigma} + \bar{0}s^{\sigma+1} \\ \bar{0} & 1 + \bar{0}s \end{pmatrix}^{\frac{1}{2}} \otimes B)$ *is the typical special automorphism of* $G^{\otimes}$ *leaving* $\mathcal{J}(\mathcal{C})$ *invariant and fixing* $A(s)$.

We now know that every automorphism of the generalized quadrangle $GQ(\mathcal{C})$ fixing (∞) and $(0,0,0)$ has the form $\theta(\sigma, A \otimes B)$, where by Theorem 2.3.6 this notation completely identifies the map $\theta(\sigma, A \otimes B)$ as an automorphism of $G^{\otimes}$. Using Theorem 2.3.4(*iii*) one can even read off $\bar{0} = (a_2/a_4)^2$ from the matrix A. This notation is not only useful for keeping track of known automorphisms, but it is useful in computing compositions of automorphisms since the following has already been established:

$$\theta(\sigma_1, A_1 \otimes B_1) \circ \theta(\sigma_2, A_2 \otimes B_2) = \theta(\sigma_1 \circ \sigma_2, A_1^{\sigma_2} A_2 \otimes B_1^{\sigma_2} B_2). \tag{2.61}$$

2.8 The Induced Oval Stabilizers

We now recall what we established in Section 1.5. We generally use the normalization as in Observation 1.6.6. $\mathcal{R}_{\alpha} = \{(\gamma \otimes \alpha, c) \in G^{\otimes} : \gamma \in F^2, c \in F\}$ is a 3-dimensional vector space over F with scalar multiplication $d(\gamma \otimes \alpha, c) = (d\gamma \otimes \alpha, d^2 c)$ and having $((1,0) \otimes \alpha, 0), ((0,1) \otimes \alpha, 0), ((0,0) \otimes \alpha, 1)$ as a basis. In fact, $a((1,0) \otimes \alpha, 0) \circ b((0,1) \otimes \alpha, 0) \circ c((0,0) \otimes \alpha, 1) = ((a,b) \otimes \alpha, c^2)$.

For $t \in \tilde{F}$, $A(t) = \{(\gamma_{yt} \otimes \beta, g(\beta, t)) \in G^{\otimes} : \beta \in F^2\}$. And $(\gamma_{yt} \otimes \beta, g(\beta, t)) \in \mathcal{R}_{\alpha}$ iff $\beta = d\alpha$ for some $d \in F$, in which case $(\gamma_{yt} \otimes \beta, g(\beta, t)) = (\gamma_{yt} \otimes d\alpha, d^2 g(\alpha, t)) = d(\gamma_{yt} \otimes \alpha, g(\alpha, t))$. So $A(t) \cap \mathcal{R}_{\alpha}$ is the point $\bar{p}_{\alpha}(t) = (\gamma_{yt} \otimes \alpha, g(\alpha, t))$ in $\overline{\mathcal{R}}_{\alpha}$. The condition K1 that $A(s) \cdot A(t) \cap A(u) = \{(0,0,0)\}$ implies that $\bar{p}_{\alpha}(s)$, $\bar{p}_{\alpha}(t)$, $\bar{p}_{\alpha}(u)$ are three noncollinear points of $\overline{\mathcal{R}}_{\alpha}$. Hence

$$\bar{\mathcal{O}}_{\alpha} = \{\bar{p}_{\alpha}(t) = (\gamma_{yt} \otimes \alpha, g(\alpha, t)) : t \in \tilde{F}\} \tag{2.62}$$

is an oval of $\overline{\mathcal{R}}_{\alpha}$ (see Proposition 1.5.2).

The mapping $\pi_{\alpha} : \overline{\mathcal{R}}_{\alpha} \to PG(2,q) : (\gamma \otimes \alpha, c) \mapsto (\gamma^{(2)}, c)$ is an isomorphism of planes mapping $\overline{\mathcal{O}}_{\alpha}$ to the oval

$$\mathcal{O}_{\alpha} = \{p_{\alpha}(t) = (\gamma_t, g(\alpha, t)) : t \in \tilde{F}\}. \tag{2.63}$$

Here if $t \in F$, $\pi_{\alpha} : (\gamma_{yt} \otimes \alpha, g(\alpha, t)) = (\gamma_{t^{\frac{1}{2}}} \otimes \alpha, g(\alpha, t)) \mapsto (\gamma_t, g(\alpha, t))$. And if $t = \infty$, $\pi_{\alpha} : (\gamma_{y\infty} \otimes \alpha, g(\alpha, \infty)) = ((0,1) \otimes \alpha, 0) \mapsto ((0,1), 0) = (\gamma_{\infty}, g(\alpha, \infty))$.

Suppose that

$$\theta = \theta(\sigma, A \otimes B) : ((\alpha, \beta), c) \mapsto ((\alpha^{\sigma}, \beta^{\sigma})(A \otimes B), \mu c^{\sigma}$$
$$+ (\alpha^{\sigma}, \beta^{\sigma})(I \otimes B) \begin{pmatrix} a_4^2 A_{(a_2/a_4)^2} & a_2 a_3 P \\ 0 & a_3^2 A_{(a_1/a_3)^2} \end{pmatrix} (I \otimes B)^T (\alpha^{\sigma}, \beta^{\sigma})^T)$$

is an automorphism of $G^{\otimes}$ leaving $\mathcal{J}(\mathcal{C})$ invariant. Here $\mathcal{C}$ is assumed to be $\frac{1}{2}$-normalized. Take $A = \left(\begin{smallmatrix} a4 & a2 \\ a3 & a1 \end{smallmatrix}\right)$, $\mu = \det(A)$, $\det(B) = 1$, and the form of θ is determined by Theorem 2.3.4. Then θ induces an isomorphism from $\overline{\mathcal{R}}_\alpha$ to $\overline{\mathcal{R}}_{\alpha^\sigma B}$ mapping $\overline{\mathcal{O}}_\alpha$ to $\overline{\mathcal{O}}_{\alpha^\sigma B}$, and hence induces an automorphism of $PG(2,q)$ mapping $\mathcal{O}_\alpha$ to $\mathcal{O}_{\alpha^\sigma B}$. Fig. 2.7 gives the maps involved.

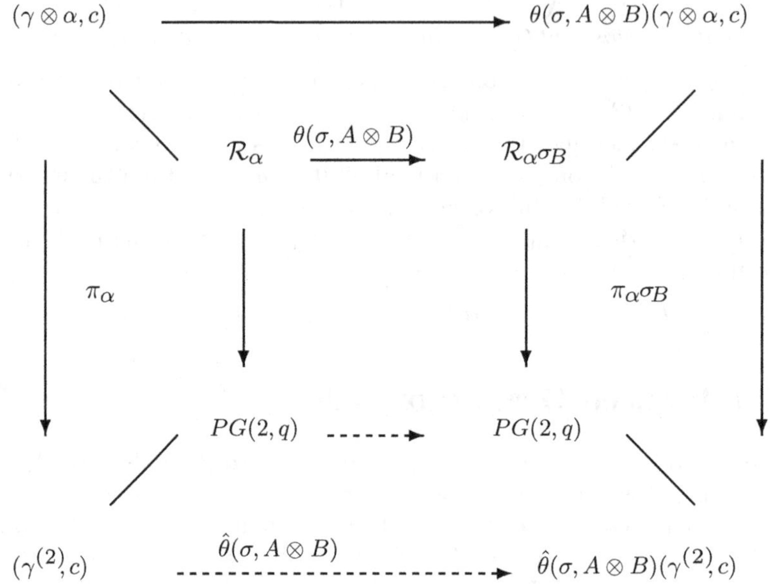

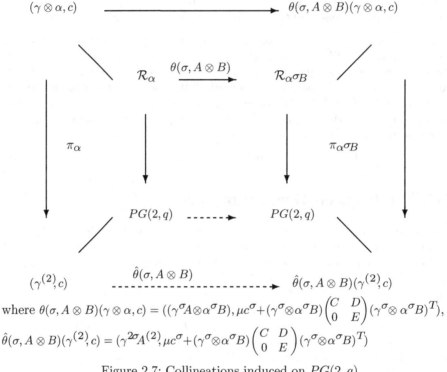

where $\theta(\sigma, A \otimes B)(\gamma \otimes \alpha, c) = ((\gamma^\sigma A \otimes \alpha^\sigma B), \mu c^\sigma + (\gamma^\sigma \otimes \alpha^\sigma B)\begin{pmatrix} C & D \\ 0 & E \end{pmatrix}(\gamma^\sigma \otimes \alpha^\sigma B)^T)$,

$\hat{\theta}(\sigma, A \otimes B)(\gamma^{(2)}, c) = (\gamma^{2\sigma} A^{(2)}, \mu c^\sigma + (\gamma^\sigma \otimes \alpha^\sigma B)\begin{pmatrix} C & D \\ 0 & E \end{pmatrix}(\gamma^\sigma \otimes \alpha^\sigma B)^T)$

Figure 2.7: Collineations induced on $PG(2,q)$

In Fig. 2.7 we have used $C = a_4^2 A_{(a2/a4)^2}$, $D = a_2 a_3 P$, and $E = a_3^2 A_{(a1/a3)^2}$. In this same context, suppose that $\alpha^\sigma B \equiv \alpha$ (i.e., $\{\alpha^\sigma B, \alpha\}$ is F-dependent). Then $\theta(\sigma, A \otimes B)$ induces an automorphism of $\overline{\mathcal{R}}_\alpha$ (resp., $PG(2,q)$) stabilizing $\overline{\mathcal{O}}_\alpha$ (resp., $\mathcal{O}_\alpha$). We shall make the appropriate specializations in Fig 2.8, giving the effect of $\hat{\theta}(\sigma, A \otimes B)$ on the ovals. For this suppose $\alpha^\sigma B = \lambda\alpha$ and see Fig. 2.8. But before considering Fig. 2.8, note the following for $t \in \tilde{F}$:

$$\begin{aligned} \hat{\theta}(\sigma, A \otimes B) : (\gamma_t, g(\alpha, t)) \mapsto {} & (\lambda^{2\sigma} \gamma_{t\sigma} A^{(2)}, \mu g(\alpha, t)^\sigma \\ & + (\gamma_{t\sigma \not{2}} \otimes \lambda\alpha) \begin{pmatrix} C & D \\ 0 & E \end{pmatrix} (\gamma_{t\sigma \not{2}} \otimes \lambda\alpha)^T) \\ = (\lambda^{2\sigma} \gamma_{t\sigma} A^{(2)}, \mu g(\alpha, t)^\sigma + \lambda^2 (\gamma_{t\sigma \not{2}} \otimes \alpha) & \begin{pmatrix} C & D \\ 0 & E \end{pmatrix} (\gamma_{t\sigma \not{2}} \otimes \alpha)^T). \end{aligned} \tag{2.64}$$

First suppose $t \in F$, so $\gamma_{t\sigma/2} \otimes \alpha = (\alpha, t^{\sigma/2}\alpha)$. Then

$$(\alpha, t^{\sigma/2}\alpha)\begin{pmatrix} C & D \\ 0 & E \end{pmatrix}(\alpha, t^{\sigma/2}\alpha)^T = \alpha C\alpha^T + \alpha(a_2a_3P)(t^{\sigma/2}\alpha)^T + t^\sigma \alpha E\alpha^T$$
$$= \alpha C\alpha^T + t^\sigma\alpha E\alpha^T = (1, t^\sigma)\begin{pmatrix} \alpha C\alpha^T \\ \alpha E\alpha^T \end{pmatrix} = \gamma_{t\sigma}(\alpha C\alpha^T, \alpha E\alpha^T)^T. \quad (2.65)$$

Next, for $t = \infty$, $\gamma_{\infty\sigma/2} \otimes \alpha = (0, \alpha)$, and $(0,\alpha)\begin{pmatrix} C & D \\ 0 & E \end{pmatrix}(0,\alpha)^T = \alpha E\alpha^T = (0,1)\begin{pmatrix} \alpha C\alpha^T \\ \alpha E\alpha^T \end{pmatrix} = \gamma_{\infty\sigma}(\alpha C\alpha^T, \alpha E\alpha^T)^T$. Now by appealing to Fig. 2.8 we have the following theorem.

Theorem 2.8.1. *With the notation developed above:*

$\hat\theta(\sigma, A\otimes B)\colon (\gamma_t, g(\alpha,t)) \mapsto (\lambda^{2\sigma}\gamma_{t\sigma}A^{(2)}, \mu g(\alpha,t)^\sigma + \lambda^2\gamma_{t\sigma}(\alpha C\alpha^T, \alpha E\alpha^T)^T)$. □

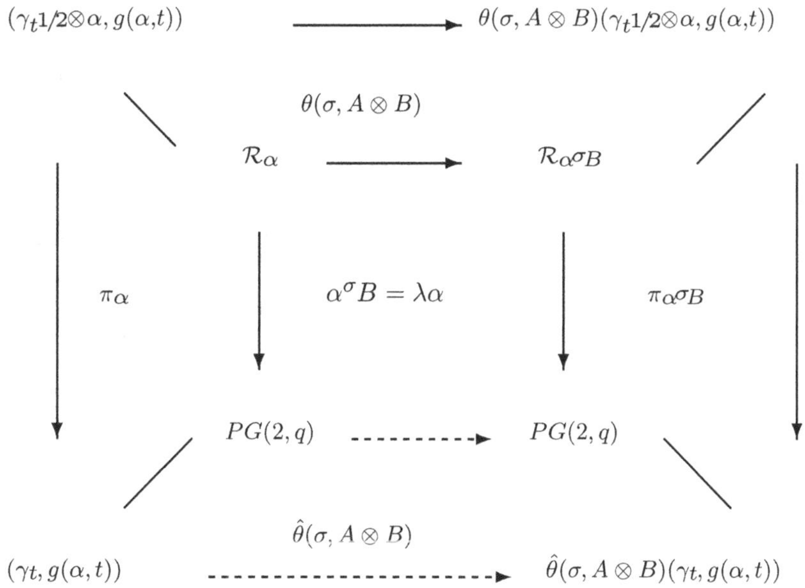

where $\theta(\sigma, A\otimes B)(\gamma_{t1/2}\otimes\alpha, g(\alpha,t)) = ((\lambda\gamma_{t\sigma/2}A\otimes\alpha), \mu g(\alpha,t)^\sigma + \lambda^2(\gamma_{t\sigma}(\alpha C\alpha^T, \alpha E\alpha^T)^T)$,
$\hat\theta(\sigma, A\otimes B)(\gamma_t, g(\alpha,t)) = (\lambda^{2\sigma}\gamma_{t\sigma}A^{(2)}, \mu g(\alpha,t)^\sigma + \lambda^2\gamma_{t\sigma}(\alpha C\alpha^T, \alpha E\alpha^T)^T)$

Figure 2.8: Induced oval stabilizer

Note. If $\theta \in \mathcal{N}$, then $\hat\theta$ fixes all points of $\mathcal{O}_\alpha$ for any $\alpha \in PG(1,q)$, so we need consider only $\theta(\sigma, A\otimes B)$ where $\det(B) = 1$.

Note. In the published version [Pa95] there is an error, viz., the λ in $\alpha^\sigma B = \lambda\alpha$ was assumed to be 1. Of course this is not always the case.

It is interesting to see what is the induced map on the herd cover. After a little computation we can write the effect of $\hat{\theta}$ on the points of an oval $\mathcal{O}_{\lambda\alpha}$ as a matrix equation. This equation will turn out to be quite useful a little later.

$$\hat{\theta}: \begin{pmatrix} 1 \\ t \\ g(\lambda\alpha, t) \end{pmatrix} \mapsto \begin{pmatrix} 1 \\ w \\ \lambda^{2\sigma} g(B^T\alpha^\sigma, w) \end{pmatrix}$$
$$= \begin{pmatrix} a & b & 0 \\ c & d & 0 \\ a\lambda^{2\sigma} g(B^T\alpha^\sigma, \frac{c}{a}) & b\lambda^{2\sigma} g(B^T\alpha^\sigma, \frac{d}{b}) & u \end{pmatrix} \begin{pmatrix} 1 \\ t \\ g(\lambda\alpha, t) \end{pmatrix}^\sigma, \quad (2.66)$$

where $w = \frac{c+dt^\sigma}{a+bt^\sigma}$.

2.9 Action of $\mathcal{H}$ on Generators of Cone K

The translation planes associated with the line spreads of $PG(3,q)$ that correspond to the flocks of the quadratic cone have derivation sets that permit the construction of other translation planes. These new planes are grouped into isomorphism classes that correspond to the orbits under the action of $\mathcal{H}$ on the generators of the quadratic cone (cf. [JL94]). Hence it is of interest in general to determine at least the number of such orbits. So we want to determine the effect of the general element $\theta = \theta(\sigma, \left(\begin{smallmatrix} 1 & 0^{\frac{1}{2}} \\ 0 & \mu \end{smallmatrix}\right) \otimes B)$ of $\mathcal{H}$, where we may assume without loss of generality that $\det(B) = 1$. Since we easily have the action of θ on the planes of $PG(3,q)$, we compute the action on the set of planes tangent to the cone. Each tangent plane passes through the point $N = (0,1,0,0)$ and through the vertex $(0,0,0,1)$.

The plane $[1,0,0,0]^T$ is tangent at the generator $\left\langle \begin{matrix} (0,0,0,1) \\ (0,0,1,0) \end{matrix} \right\rangle$.

The plane $[c^2,0,1,0]^T$ is tangent at the generator $\left\langle \begin{matrix} (0,0,0,1) \\ (1,c,c^2,0) \end{matrix} \right\rangle$.

If $B^{-1} = \left(\begin{smallmatrix} a & b \\ c & d \end{smallmatrix}\right)$, $\det B = 1$, $\sigma \in Aut(F)$, and if $\theta = \theta(\sigma, \left(\begin{smallmatrix} 1 & 0^{\frac{1}{2}} \\ 0 & \mu \end{smallmatrix}\right) \otimes B)$ is in $\mathcal{H}$, then

$$\theta: \begin{bmatrix} x \\ y \\ z \\ w \end{bmatrix} \mapsto \begin{bmatrix} \mu a^2 & \mu ab & \mu b^2 & x_0 \\ 0 & \mu & 0 & y_0 \\ \mu c^2 & \mu cd & \mu d^2 & z_0 \\ 0 & 0 & 0 & 1 \end{bmatrix} \begin{bmatrix} x^\sigma \\ y^\sigma \\ z^\sigma \\ w^\sigma \end{bmatrix}.$$

A plane $[x,y,z,w]^T$ is tangent to some generator of the cone if and only if $y = w = 0$, and scalar multiplication by a nonzero $\mu \in F$ always fixes such a plane, so we may also assume that $\mu = 1$. Then

$$\theta : [x,0,z,0]^T \mapsto [a^2x^\sigma + b^2z^\sigma, 0, c^2x^\sigma + d^2z^\sigma, 0]^T = [x', 0, z', 0]^T,$$

where

$$\left[\begin{array}{c} x' \\ z' \end{array}\right] = \left(\begin{array}{cc} a^2 & b^2 \\ c^2 & d^2 \end{array}\right) \left[\begin{array}{c} x^\sigma \\ z^\sigma \end{array}\right]. \tag{2.67}$$

This can also be represented as:

$$(x', z') = (x^\sigma, z^\sigma) \left(\begin{array}{cc} a^2 & c^2 \\ b^2 & d^2 \end{array}\right) = (x^\sigma, z^\sigma) B^{-2T}. \tag{2.68}$$

Here B^{-2T} is the result of performing three separate operations on B, viz., taking the inverse, taking the transpose, and replacing each element a by its image a^2 under the Frobenius automorphism. These three operations commute so that the map $B \mapsto B^{-2T}$ is easily seen to be a multiplicative isomorphism. Hence the action of $\mathcal{H}/\mathcal{N}$ on the generators of the cone is essentially the action of $\mathcal{H}/\mathcal{N}$ on $PG(1, q)$ given by

$$\theta(\sigma, \left(\begin{array}{cc} 1 & \bar{0}^{\frac{1}{2}} \\ 0 & \mu \end{array}\right) \otimes B) : \alpha \mapsto \alpha^\sigma B^{-2T}.$$

We denote this action by T_2.

On the other hand, the action of $\mathcal{H}/\mathcal{N}$ on the ovals $\mathcal{O}_\alpha$, $\alpha \in PG(1, q)$ is given by

$$\theta(\sigma, \left(\begin{array}{cc} 1 & \bar{0}^{\frac{1}{2}} \\ 0 & \mu \end{array}\right) \otimes B) : \alpha \mapsto \alpha^\sigma B.$$

Denote this action by T_1.

We claim that T_1 and T_2 are equivalent representations of $\mathcal{H}/\mathcal{N}$. For this purpose, define $\rho : PG(1, q) \to PG(1, q) : \alpha \mapsto \alpha^{(2)} P$. Then by a straightforward computation using $B^{-2T} = PB^{(2)}P$ we find that $\rho \circ T_2 = T_1 \circ \rho$. This result has the following interesting corollary.

Theorem 2.9.1. *The number of orbits of $\mathcal{H}/\mathcal{N}$ on the generators of the cone K is the same as the number of orbits of $\mathcal{H}/\mathcal{N}$ on the set of ovals $\mathcal{O}_\alpha$.*

Chapter 3

Aut(GQ($\mathcal{C}$))

We will denote by $GQ(\mathcal{C})$ the Generalized Quadrangle arising from the q-clan $\mathcal{C}$. In this chapter we will be concerned with the group of automorphisms of $GQ(\mathcal{C})$.

3.1 General Remarks

Recall that $\mathcal{G}$ denotes the full group of collineations of $GQ(\mathcal{C})$, and that $\mathcal{G}_0$ denotes the subgroup of $\mathcal{G}$ fixing the points $((0,0),(0,0),0)$ and (∞).

Definition 3.1.1. *If p is a point of $GQ(\mathcal{C})$, a collineation θ of $(GQ(\mathcal{C}))$ is called an* elation about p *provided the following two conditions hold:*

1. *θ fixes p linewise;*
2. *either $\theta = id$ or $\theta \neq id$ and it fixes no point of $\mathcal{P} \setminus p^{\perp}$.*

Definition 3.1.2. *A GQ is an* Elation Generalized Quadrangle (EGQ) *with base point (∞) provided there is a group G of elations about (∞) that is sharply transitive on the points of $\mathcal{P} \setminus (\infty)^{\perp}$.*

Recently, S. Payne and K. Thas [PKT02] have investigated the problem of when the set of the elations about a point p is indeed a group. They show that this is not always the case. However, if we replace condition 2. of Definition 3.1.1 with

2.′ *either $\theta = id$ or if $\theta \neq id$, then θ acts semiregularly on $\mathcal{P} \setminus p^{\perp}$,*

then the set of elations about a point p forms a group (we refer to [PKT02] for further information). A GQ arising from a q-clan $\mathcal{C}$ is an EGQ with base point (∞) having the property (G) at (∞), as we saw in Chapter 2. In the classical case, i.e., when $\mathcal{C}$ is linear and $GQ(\mathcal{C})$ is isomorphic to $H(3,q^2)$, the group $\mathcal{G}$ is transitive on points. In the non-classical case the point (∞) is fixed under $\mathcal{G}$ (see Theorem 3.1.5), but, as $GQ(\mathcal{C})$ is an EGQ, the group G of elations about (∞) is sharply transitive on the set of points not collinear with (∞). Hence in the non-classical case, to determine $\mathcal{G}$ it is sufficient to determine the group $\mathcal{G}_0$ of collineations of $GQ(\mathcal{C})$ that fix the point $(0,0,0)$ (and hence automatically fix (∞)). The F. T. provides a basic description of the elements of $\mathcal{G}_0$, but we can say quite a bit more. Our first result is from [PT91] which holds for both q even and odd.

Theorem 3.1.3. *Let $GQ(\mathcal{C})$ be the GQ arising as usual from the q-clan $\mathcal{C}$. If $GQ(\mathcal{C})$ has property (G) at some point z of $(\infty)^\perp \setminus \{(\infty)\}$, then $GQ(\mathcal{C})$ is classical.*

Proof. Recall that $\mathcal{C} = \{A_t = \left(\begin{smallmatrix} x_t & y_t \\ 0 & z_t \end{smallmatrix}\right) : t \in F\}$. We will give the proof for q even. It is possible (see Section 2.4) to assume that z is on the line $[A(\infty)]$ by applying the shifts and flips and so considering an isomorphic q-clan. Using the transitivity of $\mathcal{G}$ on the points of $[A(\infty)]$ except for (∞), we can take $z = A^*(\infty)(0,0,0)$ and by reindexing the elements of $GQ(\mathcal{C})$, we can also take $x_t = t$.

Put $\alpha = (1,0)$, $\gamma = (0,1) \in F^2$ and for each $t \in F$, put $\beta_t = (z_t, t)$. Then $\{A(\infty),\ A(\infty)(\alpha,0,0),\ A(\infty)(\gamma,0,0)\}$ is a triad of lines meeting $[A(\infty)]$ with $q+1$ transversals as in the following figure:

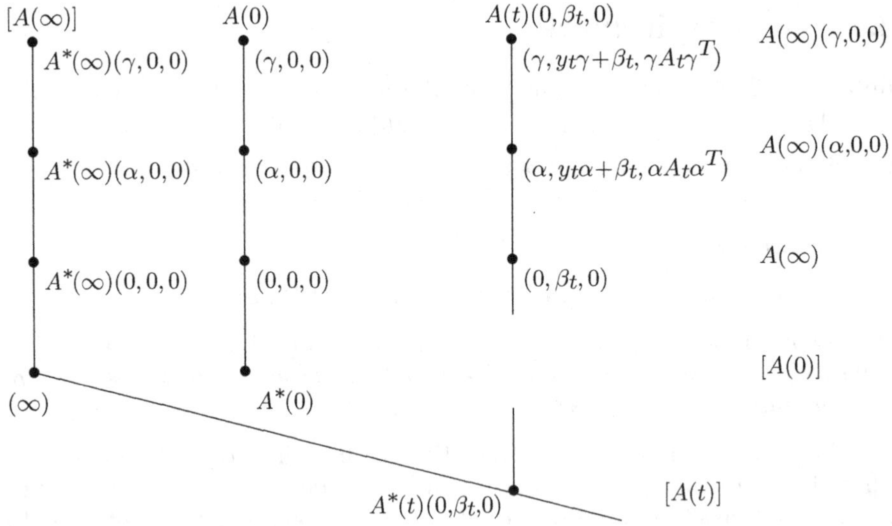

Using Property (G) at z, there must be $q-2$ values of $\delta = (d_1, d_2) \in F^2 \setminus \{(0,0), \alpha, \gamma\}$ for which some line $A(\infty)(\delta,0,0)$ through $A^*(\infty)(\delta,0,0)$ meets each of the $q+1$ transversal lines $A(t)(0,\beta_t,0)$, $t \in F$. The line $A(\infty)(\delta,0,0)$ meets $A(t)(0,\beta_t,0)$ at the point $(\delta, y_t\delta + \beta_t, \delta A_t \delta^T)$ on the line $A(\infty)(\delta, y_t\delta + \beta_t, \delta A_t\delta^T)$ which is also a line meeting $[A(\infty)]$. So for the appropriate δ all the lines of the form $A(\infty)(\delta, y_t\delta + \beta_t, \delta A_t\delta^T)$, as t varies over the elements of F, must be the same.

The above condition is equivalent to requiring that $\delta(A_t + A_s)\delta^T + (\beta_s + \beta_t)P\delta^T = 0$ for all $s, t \in F$. With $t = 0$, it is equivalent to

$$(d_1^2 + d_1)s + d_1 d_2 y_s + (d_2^2 + d_2)z_s = 0 \text{ for all } s \in F. \tag{3.1}$$

Here $d_1 \cdot d_2 \neq 0$, because if $d_2 = 0$ then Eq. (3.1) is $(d_1^2 + d_1)s = 0 \,\forall s$ which implies $d_1 = 0$ or $d_1 = 1$, which is impossible since $\delta \notin \{(0,0), \alpha\}$ and similarly $d_1 \neq 0$. Since there must be at least three choices for δ because $q \geq 8$ (for $q \leq 4$ all q-clan are classical), this forces y_s and z_s to be a constant times s. To see this take (d_1, d_2), (e_1, e_2), (f_1, f_2) as three choices for δ and put in a matrix the coefficients of Eq. (3.1):

$$M(d,e,f) = \begin{pmatrix} d_1^2 + d_1 & d_1 d_2 & d_2^2 + d_2 \\ e_1^2 + e_1 & e_1 e_2 & e_2^2 + e_2 \\ f_1^2 + f_1 & d_1 f_2 & f_2^2 + f_2 \end{pmatrix}. \tag{3.2}$$

Since for each s, (s, y_s, z_s) is in the kernel of $M(d,e,f)$, the rank of the matrix $M(d,e,f) \neq 3$. Suppose $rk(M(d,e,f)) = 1$ for all choices of three values of δ. If $d_1 = 0$, then $d_2 \in \{0,1\}$ from Eq. (3.1) which is impossible and if $d_1 = 1$, then $d_2 = 0$, which is again impossible. So for the $q-2$ elements δ the d_1's are the $q-2$ elements of $F \setminus \{0,1\}$. Then we may choose $e_1 = d_1 + 1$, with d_1 arbitrary in $F \setminus \{0,1\}$. Then Eq. (3.2) appears as

$$M(d,e,f) = \begin{pmatrix} d_1^2 + d_1 & d_1 d_2 & d_2^2 + d_2 \\ d_1^2 + d_1 & (d_1+1)e_2 & e_2^2 + e_2 \\ f_1^2 + f_1 & d_1 f_2 & f_2^2 + f_2 \end{pmatrix}. \tag{3.3}$$

Since we assume that $rk(M(d,e,f)) = 1$, it follows that $d_1 d_2 = d_1 e_2 + e_2$ and $d_2^2 + d_2 = e_2^2 + e_2$ from which we have $(d_2 + e_2)^2 = d_2 + e_2$ and hence $d_2 + e_2 \in \{0,1\}$. Now, if $d_2 = e_2$ then $e_2 = 0$ which is not possible (see above), hence $d_2 = e_2 + 1$ which implies $d_2 = d_1 + 1$. Substituting $e_1 = d_2 = d_1 + 1$ in Eq. (3.1) we get $s + y_s + z_s = 0$, so $z_s = y_s + s$. Then $z_s x_s / y_s^2 = (y_s + s)s/y_s^2$ and hence $tr(z_s x_s / y_s^2) = tr((y_s + s)s/y_s^2)$. The equation $x^2 + x + s(y_s + s)/y_s^2 = 0$ has the solution $x = s/y_s$ and so $tr((y_s + s)s/y_s^2) = 0 = tr(z_s x_s / y_s^2)$which is a contradiction because $\mathcal{C}$ is a q-clan.

So, $rk(M(d,e,f)) = 2$ and this forces the kernel to be a 1-dimensional subspace and hence there are constants a and b for which $y_s = as$ and $z_s = bs$ for all s. That is exactly the condition for the q-clan being classical. □

Lemma 3.1.4. *Let $GQ(\mathcal{C})$ be the GQ arising as usual from the q-clan $\mathcal{C}$. If P is a point not collinear with (∞) such that $P^\theta = (\infty)$ for some $\theta \in \mathcal{G}$, then the orbit $P^{\mathcal{G}}$ of P under $\mathcal{G}$ is the whole pointset $\mathcal{P}$ of $GQ(\mathcal{C})$.*

Proof. Since $GQ(\mathcal{C})$ is an EGQ, all points not collinear with (∞) are in a same orbit, so $P^{\mathcal{G}}$ consists at least of $\mathcal{P} \setminus (\infty)^{\perp}$. We claim that also the points collinear with (∞) are in $P^{\mathcal{G}}$. To this aim, take a point Q in $P^{\mathcal{G}}$ but not collinear with P and fix a line $[A(t)]$ through (∞). Since all the points not collinear with Q are in a same orbit and P is one af those, $P^{\mathcal{G}}$ contains also all the points in $[A(t)]$ except for one (let's call it P_1, that is the unique point obtained as intersection of $[A(t)]$ and the unique line through Q meeting $[A(t)]$). Now, if we take a point not collinear with Q and not collinear with P_1, it follows in the same way as before that Q and P_1 are in a same orbit, hence $P_1 \in P^{\mathcal{G}}$. Doing the same for all the lines through (∞), it follows that $P^{\mathcal{G}}$ contains also all the points collinear with (∞), hence $P^{\mathcal{G}} = \mathcal{P}$. □

Theorem 3.1.5. *Let $\mathcal{C}$ be a non-classical q-clan. Then the point (∞) is fixed under the complete group $\mathcal{G}$ of collineations of $GQ(\mathcal{C})$.*

Proof. Suppose (∞) is not fixed under the complete group $\mathcal{G}$ of collineations of $GQ(\mathcal{C})$. Then there is a point $P \neq (\infty)$ such that $(\infty)^{\theta} = P$, where θ is a collineation of $GQ(\mathcal{C})$. If P is collinear with (∞), it is a point distinct from (∞), collinear with (∞) and having property (G), which is not possible by Theorem 3.1.3. If P is not collinear with (∞), then by Lemma 3.1.4, the orbit of P under $\mathcal{G}$ (or equivalently, the orbit of (∞) under $\mathcal{G}$) is the whole pointset of $GQ(\mathcal{C})$. So we can find a point collinear with (∞) with the property (G). This is again a contradiction by Theorem 3.1.3. Hence the point (∞) must be fixed by $\mathcal{G}$ and the theorem follows. □

3.2 An involution of GQ($\mathcal{C}$)

For $q = 2^e$ there are five known infinite families of q-clans. In the Subiaco Notebook there is a chapter devoted to the three families of monomial q-clans and a longer chapter devoted to the Subiaco examples. However, with the discovery in [COP03] of a unified construction of the so-called cyclic examples that includes a new family, it now seems more appropriate to treat the cyclic examples together and the examples of S. E. Payne (from [Pa85]) separately. Moreover, our unified treatment of the cyclic GQ can now be very much shorter than previous treatments of the Subiaco GQ. In any case the following theorem will apply to all five families.

Theorem 3.2.1. *Suppose $\mathcal{C} = \{A_t = \begin{pmatrix} f(t) & t^{\frac{1}{2}} \\ 0 & g(t) \end{pmatrix} : t \in F\}$ is a q-clan for functions $f(t)$ and $g(t)$ satisfying $f(t^{-1}) = t^{-1}g(t)$ (equivalently, $g(t^{-1}) = t^{-1}f(t)$). Then $\varphi = \theta(id, P \otimes P) : (\alpha, \beta, c) \mapsto (\beta P, \alpha P, c + \alpha \circ \beta)$ is an automorphism of $GQ(\mathcal{C})$ interchanging $A(t)$ and $A(t^{-1})$ for all $t \in \tilde{F}$.*

Proof. For all $t \in \tilde{F}$, $A(t) = \{(\gamma_{yt} \otimes \alpha, g(\alpha, t)) : \alpha \in F^2\}$. (Recall that $\gamma_{y\infty} = (0, 1)$.) And $\theta(id, P \otimes P) : (\gamma_{yt} \otimes \alpha, g(\alpha, t)) \mapsto (\gamma_{yt}P \otimes \alpha P, g(\alpha, t) + 0)$. Clearly $\gamma_{y\infty}P = \gamma_{y0}$ and $\gamma_{y0}P = \gamma_{y\infty}$ Also $g(\alpha, 0) = g(\alpha, \infty) = 0$. So $\theta(id, P \otimes P)$

interchanges $A(0)$ and $A(\infty)$. So suppose $0 \neq t \in F$. Then $\gamma_{yt}P = (1, t^{\frac{1}{2}})P = t^{\frac{1}{2}}(1, t^{-\frac{1}{2}}) = t^{\frac{1}{2}}\gamma_{y_{t^{-1}}}$. Hence

$$\begin{aligned}(\gamma_{yt}P \otimes \alpha P, g(\alpha, t)) &= (\gamma_{y_{t^{-1}}} \otimes t^{\frac{1}{2}}\alpha P, t^{-1}(t^{\frac{1}{2}}\alpha P \cdot PA_tP \cdot (t^{\frac{1}{2}}\alpha P)^T)) \\ &\equiv (\gamma_{y_{t^{-1}}} \otimes t^{\frac{1}{2}}\alpha P, (t^{\frac{1}{2}}\alpha P)\begin{pmatrix} t^{-1}g(t) & t^{-\frac{1}{2}} \\ 0 & t^{-1}f(t) \end{pmatrix}(t^{\frac{1}{2}}\alpha P)^T) \\ &= (\gamma_{y_{t^{-1}}} \otimes t^{\frac{1}{2}}\alpha P, g(t^{\frac{1}{2}}\alpha P, t^{-1})) \in A(t^{-1}). \quad \square\end{aligned}$$

We will continue to use the symbol φ to denote the special involution (also sometimes called the "flip") described in the preceding theorem. However, later when we have a unique involution I_t fixing the line $[A(t)]$ for each $t \in \tilde{F}$, φ will be denoted by I_1.

3.3 The Automorphism Group of the Herd Cover

Recall all the definitions and notation of Chapter 1 Section 1.6. Since herd covers and q-clans are equivalent objects in some sense, (see Theorem 1.7.1), there ought to be a close connection between the automorphism group of the herd cover and the automorphism group of the associated GQ. Since the members of a herd cover are the projections of the ovals in $\bar{\mathcal{R}}_\alpha$ which live in $G^{\otimes}$, there also ought to be a direct connection with $\mathrm{Aut}(G^{\otimes})$. We propose the following.

Definition 3.3.1. *The* automorphism group G_0 of the herd cover $\mathcal{H}(\mathcal{C})$ *is the subgroup of $P\Gamma L(3, q)$ that induces a permutation of the oval covers of $\mathcal{H}(\mathcal{C})$.*

According to this definition, the automorphism group of a herd of ovals would naturally be the group induced on a herd of ovals by the automorphism group of the corresponding herd cover. We also note that the induced group of the herd of ovals depends on the normalization used to define the herd of ovals.

Definition 3.3.2. *The* automorphism group $\bar{G}_0$ of the profile $\overline{\mathcal{H}(\mathcal{C})}$ *of the herd cover is the subgroup of $Aut(G^{\otimes})$ that induces a permutation of the ovals of $\overline{\mathcal{H}(\mathcal{C})}$.*

We observe that the automorphism group of a herd cover is isomorphic to the automorphism group of the profile of that herd cover.

Put

$$T = \{(\gamma \otimes \alpha, c) \in G^{\otimes}; \gamma, \alpha \in F^2,\ c \in F\}.$$

Clearly

$$T = \cup\{\mathcal{L}_\gamma : \gamma \in PG(1, q)\} \cup \{\mathcal{R}_\alpha : \alpha \in PG(1, q)\}.$$

It is easy to show that any subgroup of $G^{\otimes}$ spanned by any two distinct "points" of T and lying entirely in T must either lie in some $\mathcal{L}_\gamma$ or in some $\mathcal{R}_\alpha$. Let p_1 and p_2 be any two distinct o-points lying in some $\mathcal{L}_\gamma$. Then $p_1 \cdot p_2 = p_3$ also lies in $\mathcal{L}_\gamma$.

Hence any $\theta \in \mathrm{Aut}(G^{\otimes})$ that induces a permutation of the ovals maps all three o-points p_1, p_2, p_3 to o-points for which the third is the "sum" of the first two. Hence the three images also lie in a same $\mathcal{L}_{\gamma'}$. This implies that θ must also permute the $\mathcal{L}_\gamma$ among themselves. Also, the points of T that are contained in the ovals and lie in some $\mathcal{L}_\gamma$ lie on the line that is the member of the 4-gonal family in $\mathcal{L}_\gamma$. Hence θ must permute the members of the 4-gonal family, i.e., θ induces a collineation of the GQ $GQ(\mathcal{C})$. As θ clearly fixes $(0,0,0)$, the Fundamental Theorem tells us what form θ must have. We have shown that

Theorem 3.3.3. *The automorphism group* G_0 $(\bar{G}_0)$ *of the (profile of the) herd cover* $\mathcal{H}(\mathcal{C})$ *is isomorphic to the group of automorphisms of* $GQ(\mathcal{C})$ *that fix the points* (∞) *and* $(0,0,0)$.

Note. In the above argument we only used the fact that $\theta \in \bar{G}_0$ permutes the o-points among themselves. But of course since such a θ induces an automorphism of the GQ, it will have to permute the ovals among themselves.

Let $\theta = \theta(\sigma, A \otimes B) \in G_0$. Suppose $A = \begin{pmatrix} a4 & a2 \\ a3 & a1 \end{pmatrix} = \begin{pmatrix} a^{\frac12} & c^{\frac12} \\ b^{\frac12} & d^{\frac12} \end{pmatrix}$, $u = \det(A)$, $1 = \det(B)$. According to Theorem 2.3.4 the action of θ on the points of T is given by

$$\theta = \theta(\sigma, A \otimes B) : (\gamma \otimes \alpha, r) \mapsto$$
$$(A^T\gamma^\sigma \otimes B^T\alpha^\sigma, ur^\sigma + \gamma^{2\sigma}((\alpha^\sigma)^T BCB^T\alpha^\sigma, (\alpha^\sigma)^T BEB^T\alpha^\sigma)^T), \quad (3.4)$$

where $C = a_4^2 A_{\left(\frac{\alpha 2}{\alpha 4}\right)^2} = aA_{\left(\frac{c}{a}\right)}$ and $E a_3^2 A_{\left(\frac{\alpha 1}{\alpha 3}\right)^2} = bA_{\left(\frac{d}{b}\right)}$.

As we saw in Chapter 2, Section 2.8, θ induces a map $\hat{\theta} = \hat{\theta}(\sigma, A \otimes B)$ on the herd cover. We recall here the form of $\hat{\theta}$ we have already established:

$$\hat{\theta}: \begin{pmatrix} 1 \\ t \\ g(\lambda\alpha, t) \end{pmatrix} \mapsto \begin{pmatrix} 1 \\ w \\ \lambda^{2\sigma} g(B^T\alpha^\sigma, w) \end{pmatrix}$$
$$= \begin{pmatrix} a & b & 0 \\ c & d & 0 \\ a\lambda^{2\sigma} g\left(B^T\alpha^\sigma, \frac{c}{a}\right) & b\lambda^{2\sigma} g\left(B^T\alpha^\sigma, \frac{d}{b}\right) & u \end{pmatrix} \begin{pmatrix} 1 \\ t \\ g(\lambda\alpha, t) \end{pmatrix}^\sigma, \quad (3.5)$$

where $w = \frac{c + dt^\sigma}{a + bt^\sigma}$.

And, as far as a normalized herd of ovals is concerned, we get

$$\hat{\theta}(\sigma, A \otimes B) : (\gamma_t, g(\alpha, t)) \mapsto (\lambda^{2\sigma}\gamma_{t\sigma} A^{(2)}, \mu g(\alpha, t)^\sigma$$
$$+ (\gamma_{t\sigma} \not{2} \otimes \lambda\alpha) \begin{pmatrix} C & D \\ 0 & E \end{pmatrix} (\gamma_{t\sigma} \not{2} \otimes \lambda\alpha)^T)$$
$$= (\lambda^{2\sigma}\gamma_{t\sigma} A^{(2)}, \mu g(\alpha, t)^\sigma + \lambda^2 (\gamma_{t\sigma} \not{2} \otimes \alpha) \begin{pmatrix} C & D \\ 0 & E \end{pmatrix} (\gamma_{t\sigma} \not{2} \otimes \alpha)^T). \quad (3.6)$$

3.4 The Magic Action of O'Keefe and Penttila

Let $\mathcal{F} = \{f : F \to F : f(0) = 0\}$. Each element of $\mathcal{F}$ can be expressed as a polynomial in one variable of degree at most $q-1$, and $\mathcal{F}$ is naturally a vector space over F. For $f(t) = \sum a_i t^i \in \mathcal{F}$ and $\sigma \in Aut(F)$, put $f^\sigma(t) = \sum a_i^\sigma t^i = (f(t^{\frac{1}{\sigma}}))^\sigma$. Start with the group $P\Gamma L(2,q)$ acting on the projective line $PG(1,q)$,

$$P\Gamma L(2,q) = \{x \mapsto Ax^\sigma\colon A \in GL(2,q),\ \sigma \in Aut(F)\}.$$

(Clearly $x \in F^2$ is written as a column vector.)

We are going to construct an action $\mathcal{M}\colon \Gamma L(2,q) \to Sym_{\mathcal{F}}$ of $\Gamma L(2,q)$ on the set $\mathcal{F}$. For each $f \in \mathcal{F}$ and each $\psi \in \Gamma L(2,q)$, where $\psi\colon x \mapsto Ax^\sigma$ for $A = \left(\begin{smallmatrix} a & b \\ c & d\end{smallmatrix}\right) \in GL(2,q)$ and $\sigma \in Aut(F)$, let the image of f under $\mathcal{M}(\psi)$ be the function $\psi f : F \to F$ such that

$$\psi f(t) = |A|^{-\frac{1}{2}}\left\{(bt+d)f^\sigma\left(\frac{at+c}{bt+d}\right) + btf^\sigma\left(\frac{a}{b}\right) + df^\sigma\left(\frac{c}{d}\right)\right\}. \tag{3.7}$$

Lemma 3.4.1. *$\mathcal{M}$ is an action (called the* magic action*) of $\Gamma L(2,q)$ on the set $\mathcal{F}$ whose kernel contains (so equals) the set of scalar matrices $aI : 0 \neq a \in F$ that form the center of $\Gamma L(2,q)$. (The induced action of $P\Gamma L(2,q)$ is also called the magic action.)*

Proof. If f is an element in $\mathcal{F}$, $\psi f(0) = |A|^{-\frac{1}{2}}df^\sigma(\frac{c}{d}) + |A|^{-\frac{1}{2}}df^\sigma(\frac{c}{d}) = 0$, as q is even, hence $\psi f \in \mathcal{F}$. We show now that $\mathcal{M}$ yields an action of $\Gamma L(2,q)$ on $\mathcal{F}$, i.e., $\mathcal{M}(\psi_2\psi_1) = \mathcal{M}(\psi_2) \circ \mathcal{M}(\psi_1)$ for any $\psi_1, \psi_2 \in \Gamma L(2,q)$ and $1f(t) = f(t)$. The identity element of $\Gamma L(2,q)$ is $1 : x \mapsto Ix$, where I is the identity matrix, so $1f(t) = f(t)$. Let $\psi_1 : x \mapsto A_1 x^{\sigma 1}$ and $\psi_2 : x \mapsto A_2 x^{\sigma 2}$, where $A_1 = \left(\begin{smallmatrix} a1 & b1 \\ c1 & d1\end{smallmatrix}\right)$ and $A_2 = \left(\begin{smallmatrix} a2 & b2 \\ c2 & d2\end{smallmatrix}\right)$ are in $GL(2,q)$ and $\sigma_1, \sigma_2 \in Aut(GF(q))$. Then (applying ψ_1 first), $\psi_2\psi_1 : x \mapsto A_2 A_1^{\sigma 2} x^{\sigma 1 \sigma 2}$ so that, since

$$A_2 A_1^{\sigma 2} = \begin{pmatrix} A_{12} & B_{12} \\ C_{12} & D_{12}\end{pmatrix} = \begin{pmatrix} a_1^{\sigma 2}a_2 + c_1^{\sigma 2}b_2 & b_1^{\sigma 2}a_2 + d_1^{\sigma 2}b_2 \\ a_1^{\sigma 2}c_2 + c_1^{\sigma 2}d_2 & b_1^{\sigma 2}c_2 + d_1^{\sigma 2}d_2\end{pmatrix},$$

then the action of $\psi_2\psi_1$ on $\mathcal{F}$ is:

$$(\psi_2\psi_1)f(t) = |A_2 A_1^{\sigma 2}|^{-1/2}\left[(B_{12}t + D_{12})f^{\sigma 1\sigma 2}\left(\frac{A_{12}t + C_{12}}{B_{12}t + D_{12}}\right)\right.$$
$$\left. + B_{12}tf^{\sigma 1\sigma 2}\left(\frac{A_{12}}{B_{12}}\right) + D_{12}f^{\sigma 1\sigma 2}\left(\frac{C_{12}}{D_{12}}\right)\right].$$

On the other hand,

$$\psi_1 f(y) = |A_1|^{-\frac{1}{2}}(b_1 y + d_1)\left[f\left(\frac{a_1^{1/\sigma 1}y^{1/\sigma 1} + c_1^{1/\sigma 1}}{b_1^{1/\sigma 1}y^{1/\sigma 1} + d_1^{1/\sigma 1}}\right)\right]^{\sigma 1}$$
$$+ |A_1|^{-\frac{1}{2}}b_1\left[f\left(\frac{a_1^{1/\sigma 1}}{b_1^{1/\sigma 1}}\right)\right]^{\sigma 1} y + |A_1|^{-\frac{1}{2}}d_1\left[f\left(\frac{c_1^{1/\sigma 1}}{d_1^{1/\sigma 1}}\right)\right]^{\sigma 1}$$

so that

$$
\begin{aligned}
\psi_2(\psi_1 f)(t) &= |A_2|^{-1/2}(b_2 t + d_2)\left[\psi_1 f\left(\frac{a_2^{1/\sigma 2} t^{1/\sigma 2} + c_2^{1/\sigma 2}}{b_2^{1/\sigma 2} t^{1/\sigma 2} + d_2^{1/\sigma 2}}\right)\right]^{\sigma 2} \\
&\quad + |A_2|^{-1/2} b_2 \left[\psi_1 f\left(\frac{a_2^{1/\sigma 2}}{b_2^{1/\sigma 2}}\right)\right]^{\sigma 2} t + |A_2|^{-\frac{1}{2}} d_2 \left[\psi_1 f\left(\frac{c_2^{1/\sigma 2}}{d_2^{1/\sigma 2}}\right)\right]^{\sigma 2} \\
&= |A_2|^{-1/2}(b_2 t + d_2)\left[|A_1|^{-\frac{1}{2}}\left(b_1\left(\frac{a_2^{1/\sigma 2} x^{1/\sigma 2} + c_2^{1/\sigma 2}}{b_2^{1/\sigma 2} x^{1/\sigma 2} + d_2^{1/\sigma 2}}\right) + d_1\right)\right. \\
&\quad \times \left[f\left(\frac{a_1^{1/\sigma 1}\left(\frac{a_2^{1/\sigma 2} t^{1/\sigma 2} + c_2^{1/\sigma 2}}{b_2^{1/\sigma 2} t^{1/\sigma 2} + d_2^{1/\sigma 2}}\right)^{1/\sigma 1} + c_1^{1/\sigma 1}}{b_1^{1/\sigma 1}\left(\frac{a_2^{1/\sigma 2} t^{1/\sigma 2} + c_2^{1/\sigma 2}}{b_2^{1/\sigma 2} t^{1/\sigma 2} + d_2^{1/\sigma 2}}\right)^{1/\sigma 1} + d_1^{1/\sigma 1}}\right)\right]^{\sigma 1} \\
&\quad \left. + |A_1|^{-1/2} b_1 \left[f\left(\frac{a_1^{1/\sigma 1}}{b_1^{1/\sigma 1}}\right)\right]^{\sigma 1}\left(\frac{a_2^{1/\sigma 2} t^{1/\sigma 2} + c_2^{1/\sigma 2}}{b_2^{1/\sigma 2} t^{1/\sigma 2} + d_2^{1/\sigma 2}}\right) + |A_1|^{-1/2} d_1 \left[f\left(\frac{c_1^{1/\sigma 1}}{d_1^{1/\sigma 1}}\right)\right]^{\sigma 1}\right]^{\sigma 2} \\
&\quad + |A_2|^{-1/2} b_2 \left[|A_1|^{-1/2}\left(b_1 \frac{a_2^{1/\sigma 2}}{b_2^{1/\sigma 2}} + d_1\right)\left[f\left(\frac{a_1^{1/\sigma 1}\frac{a_2^{1/\sigma 1\sigma 2}}{b_2^{1/\sigma 1\sigma 2}} + c_1^{1/\sigma 1}}{b_1^{1/\sigma 1}\frac{a_2^{1/\sigma 1\sigma 2}}{b_2^{1/\sigma 1\sigma 2}} + d_1^{1/\sigma 1}}\right)\right]^{\sigma 1}\right. \\
&\quad \left. + |A_1|^{-1/2} b_1 \left[f\left(\frac{a_1^{1/\sigma 1}}{b_1^{1/\sigma 1}}\right)\right]^{\sigma 1} \frac{a_2^{1/\sigma 2}}{b_2^{1/\sigma 2}} + |A_1|^{-1/2} d_1 \left[f\left(\frac{c_1^{1/\sigma 1}}{d_1^{1/\sigma 1}}\right)\right]^{\sigma 1}\right]^{\sigma 2} t \\
&\quad + |A_2|^{-1/2} d_2 \left[|A_1|^{-1/2}\left(b_1 \frac{c_2^{1/\sigma 2}}{d_2^{1/\sigma 2}} + d_1\right)\left[f\left(\frac{a_1^{1/\sigma 1}\frac{c_2^{1/\sigma 1\sigma 2}}{d_2^{1/\sigma 1\sigma 2}} + c_1^{1/\sigma 1}}{b_1^{1/\sigma 1}\frac{c_2^{1/\sigma 1\sigma 2}}{d_2^{1/\sigma 1\sigma 2}} + d_1^{1/\sigma 1}}\right)\right]^{\sigma 1}\right. \\
&\quad \left. + |A_1|^{-1/2} b_1 \left[f\left(\frac{a_1^{1/\sigma 1}}{b_1^{1/\sigma 1}}\right)\right]^{\sigma 1} \frac{c_2^{1/\sigma 2}}{d_2^{1/\sigma 2}} + |A_1|^{-1/2} d_1 \left[f\left(\frac{c_1^{1/\sigma 1}}{d_1^{1/\sigma 1}}\right)\right]^{\sigma 1}\right]^{\sigma 2}
\end{aligned}
$$

which turns out to be equal to $(\psi_2\psi_1)f(t)$. Thus $\mathcal{M}(\psi_2\psi_1) = \mathcal{M}(\psi_2) \circ \mathcal{M}(\psi_1)$, hence $\mathcal{M}$ defines an action of $\Gamma L(2,q)$ on $\mathcal{F}$. Let $\psi \in \Gamma L(2,q)$ be $\psi : x \mapsto aIx$ where I is the identity matrix and $a \neq 0$, then

$$\psi f(t) = (a^2)^{-1/2} a f\left(\frac{at}{a}\right) = f(t).$$

Hence, $\mathcal{M}$ induces an action of $P\Gamma L(2,q)$ on $\mathcal{F}$ because the elements of $\Gamma L(2,q)$ corresponding to the scalar matrices (i.e., the center of $P\Gamma L(2,q)$) are in the kernel of $\mathcal{M}$. □

Lemma 3.4.2. *The function $f \in \mathcal{F}$ defined by $f(t) = t^{\frac{1}{2}}$ is fixed by every $\psi \in P\Gamma L(2,q)$ under the magic action.*

Proof. Let $\psi : x \mapsto Ax^\sigma$, where $A = \left(\begin{smallmatrix} a & b \\ c & d \end{smallmatrix}\right) \in GL(2,q)$ and $\sigma \in Aut(GF(q))$. Note that if $f(t) = t^{1/2}$, then $f^\sigma(t) = f(t)$ and

$$\psi f(t) = \psi(t^{1/2}) = |A|^{-1/2}\left((bt+d)\left(\frac{at+c}{bt+d}\right)^{1/2} + bt\left(\frac{a}{b}\right)^{1/2} + d\left(\frac{c}{d}\right)^{1/2}\right)$$
$$= |A|^{-1/2}(a^{1/2}d^{1/2}t^{1/2} + b^{1/2}c^{1/2}t^{1/2}) = t^{1/2}.$$

Hence the lemma follows. □

Lemma 3.4.3. *The magic action of $P\Gamma L(2,q)$ on $\mathcal{F}$ is (projective) semi-linear and the magic action of the subgroup $PGL(2,q)$ is (projective) linear.*

Proof. Let $\psi \in P\Gamma L(2,q)$ be $\psi : x \mapsto Ax^\sigma$, where $A = \left(\begin{smallmatrix} a & b \\ c & d \end{smallmatrix}\right) \in GL(2,q)$ and $\sigma \in Aut(GF(q))$. Let f, $g \in \mathcal{F}$ and $k \in GF(q)$. Since σ is an automorphism of the field, $(f+g)^\sigma = f^\sigma + g^\sigma$ and $(kf)^\sigma = k^\sigma f^\sigma$, then the theorem follows by straightforward verification. □

It is quite useful to have the following list of generators of $P\Gamma L(2,q)$ and their magic actions on $\mathcal{F}$.

$$\sigma_a : x \mapsto \begin{pmatrix} a & 0 \\ 0 & 1 \end{pmatrix} x, \quad \sigma_a f(t) = a^{-\frac{1}{2}} f(at), \text{ for } 0 \neq a \in F.$$

$$\tau_c : x \mapsto \begin{pmatrix} 1 & 0 \\ c & 1 \end{pmatrix} x, \quad \tau_c f(t) = f(t+c) + f(c), \text{ for } c \in F.$$

$$\phi : x \mapsto \begin{pmatrix} 0 & 1 \\ 1 & 0 \end{pmatrix} x, \quad \phi f(t) = t f(t^{-1}).$$

$$\rho_\sigma : x \mapsto x^\sigma, \quad \rho_\sigma f(t) = f^\sigma(t).$$

O'Keefe and Penttila [OP02] proceed to use the magic action in the study of ovals and herds of ovals. However, if we shift to the examination of oval covers and herd covers, the $\mathcal{F}$ setting for the magic action is not optimal. Since $\mathcal{F}$ is a vector space over F, we can pass to the projective space $\hat{\mathcal{F}}$ whose points $\langle f \rangle$ are the 1-dimensional subspaces of $\mathcal{F}$ spanned by a nonzero $f \in \mathcal{F}$. It is easy to see that for all $\lambda \in F^*$, $\psi(\lambda f) = \lambda^\sigma \psi f$, and so, the magic action can be lifted to $\hat{\mathcal{F}}$. By a fairly common abuse of notation we will continue to use ψ to denote the result of the magic action on the points of $\hat{\mathcal{F}}$, i.e., $\psi\langle f \rangle := \langle \psi f \rangle$. We will now rephrase further results of [OP02] in the $\hat{\mathcal{F}}$ setting. We will use the following notation. Let $\mathcal{D}(f) = \{(1,t,f(t)) \colon t \in F\} \cup \{(0,1,0)\}$ for any o-permutation f. That is, $\mathcal{D}(f)$ is an oval in $PG(2,q)$. For any primitive element ζ of F define

$$\mathcal{D}_\zeta(f) = \{(1,t,f(t),\zeta f(t),\zeta^2 f(t),\ldots,\zeta^{q-2} f(t)) \colon t \in F\} \cup \{(0,1,0,\ldots,0)\} \quad (3.8)$$

to be a set of points in $PG(q,q)$ for any o-permutation f. $\mathcal{D}_\zeta(f)$ is a coordinate representation of an oval cover. The oval cover can be recovered by projecting the coordinates x_0, x_1, x_i of $\mathcal{D}_\zeta(f)$ into the same $PG(2,q)$ for each i, $2 \leq i \leq q$. Notice that if we rearrange, in any way, the last $q-1$ coordinates of $\mathcal{D}_\zeta(f)$, these projections will give the same oval cover. We should therefore consider two of these sets to be equivalent if one can be obtained from the other by such a rearrangement. The equivalence classes may be described by

$$\mathcal{D}(\langle f\rangle) = \{(1, t, \overline{f(t)}B) \colon B \text{ is a permutation matrix}, t \in F\},$$

where $\overline{f(t)} = (f(t), \zeta f(t), \zeta^2 f(t), \ldots, \zeta^{q-2} f(t))$ for any fixed primitive element $\zeta \in F$. Note that, as a special case, $\mathcal{D}_\zeta(f)$ and $\mathcal{D}_\eta(f)$ will be equivalent for different primitive elements ζ and η.

The next theorem is what makes the magic action useful to us.

Theorem 3.4.4. *Let $f \in \mathcal{F}$ be an o-permutation for $PG(2,q)$ and let $\psi \in P\Gamma L(2,q)$ be $\psi \colon x \mapsto Ax^\sigma$ for $A = \left(\begin{smallmatrix} a & b \\ c & d \end{smallmatrix}\right) \in GL(2,q)$ and $\sigma \in Aut(F)$. Then $g \in \psi\langle f\rangle$ is also an o-permutation. In fact, for any o-permutation f, $\mathcal{D}(\psi f) = \bar{\psi}_f \mathcal{D}(f)$ where $\bar{\psi}_f \colon x \mapsto \bar{A}_f x^\sigma$ $(x \in F^3)$ for*

$$\bar{A}_f = \begin{pmatrix} a & b & 0 \\ c & d & 0 \\ a\psi f\left(\frac{c}{a}\right) & b\psi f\left(\frac{d}{b}\right) & |A|^{\frac{1}{2}} \end{pmatrix}. \tag{3.9}$$

In terms of oval covers, we have $\mathcal{D}_{\zeta^\sigma}(\psi f) = \bar{\psi}_{\langle f\rangle} \mathcal{D}_\zeta(f)$ where $\bar{\psi}_{\langle f\rangle} \in P\Gamma L(q+1,q)$ is such that $\bar{\psi}_{\langle f\rangle} \colon x \mapsto \bar{A}_{\langle f\rangle} x^\sigma$ $(x \in F^{q+1})$ for

$$\bar{A}_{\langle f\rangle} = \begin{pmatrix} A & \mathbf{0}_{2\times q-1} \\ D & |A|^{\frac{1}{2}} \mathbf{I}_{q-1\times q-1} \end{pmatrix} \tag{3.10}$$

where

$$D = \begin{pmatrix} a\psi f\left(\frac{c}{a}\right) & b\psi f\left(\frac{d}{b}\right) \\ a\psi \zeta f\left(\frac{c}{a}\right) & b\psi \zeta f\left(\frac{d}{b}\right) \\ a\psi \zeta^2 f\left(\frac{c}{a}\right) & b\psi \zeta^2 f\left(\frac{d}{b}\right) \\ \vdots & \vdots \\ a\psi \zeta^{q-2} f\left(\frac{c}{a}\right) & b\psi \zeta^{q-2} f\left(\frac{d}{b}\right) \end{pmatrix}.$$

Proof. We verify that $\mathcal{D}(\psi f) = \bar{\psi}_f \mathcal{D}(f)$ by direct computation. First,

$$\begin{aligned}\bar{\psi}_f \mathcal{D}(f) &= \{\bar{\psi}_f(1,t,f(t)) : t \in GF(q)\} \cup \{\bar{\psi}_f(0,1,0)\} \\ &= \{\bar{A}_f(1,t^\sigma,(f(t))^\sigma) : t \in GF(q)\} \cup \{\bar{A}_f(0,1,0)\} \\ &= \{\bar{A}_f(1,t,f^\sigma(t)) : t \in GF(q)\} \cup \{\bar{A}_f(0,1,0)\}\end{aligned}$$

$$= \left\{\left(a+bt, c+dt, |A|^{1/2} f^\sigma(t) + b\psi f\left(\frac{d}{b}\right) t + a\psi f\left(\frac{c}{a}\right)\right) : t \in GF(q)\right\}$$
$$\cup \left\{\left(b, d, b\psi f\left(\frac{d}{b}\right)\right)\right\}$$

$$= \left\{\left(1, \frac{c+dt}{a+bt}, \frac{|A|^{1/2} f^\sigma(t) + b\psi f\left(\frac{d}{b}\right) t + a\psi f\left(\frac{c}{a}\right)}{a+bt}\right) : t \in GF(q) \setminus \left\{\frac{a}{b}\right\}\right\}$$
$$\cup \left\{(0,1,0), \left(b, d, b\psi f\left(\frac{d}{b}\right)\right)\right\}.$$

On substituting $s = (c+dt)/(a+bt)$ (hence $t = (as+c)/(bs+d)$), we obtain

$$\left\{(1,s,r) : s \in GF(q) \setminus \left\{\frac{d}{b}\right\}\right\} \cup \left\{(0,1,0), \left(b, d, b\psi f\left(\frac{d}{b}\right)\right)\right\},$$

where

$$r = \frac{|A|^{1/2} f^\sigma\left(\frac{as+c}{bs+d}\right)(bs+d) + b\psi f\left(\frac{d}{b}\right)(as+c) + a\psi f\left(\frac{c}{a}\right)(bs+d)}{|A|},$$

which is in fact $\{(1,s,\psi f(s)) : s \in GF(q)\} \cup \{(0,1,0)\}$. This set of points is an oval, containing the points $(1,0,0)$ and $(0,1,0)$, and with nucleus $(0,0,1)$; so ψf is an o-permutation.

We consider now the oval cover extension of this result. By the first part of the theorem, for ζ a primitive element of F and $0 \le i < q-1$, $\mathcal{D}(\psi\zeta^i f) = \bar{\psi}_{\zeta^i f}\mathcal{D}(\zeta^i f)$. Since $\mathcal{D}(\psi\zeta^i f) = \mathcal{D}(\zeta^{\sigma i}\psi f)$ the result follows. □

Corollary 3.4.5. *Let $f \in \mathcal{F}$ be an o-permutation, and let $\psi \in P\Gamma L(2,q)$. If $\psi f \in \langle f\rangle$, then $\bar{\psi}_f$ multiplied by $\left(\begin{smallmatrix}1&0&0\\0&1&0\\0&0&k\end{smallmatrix}\right)$ where $k \in F^*$, is in the stabilizer of $\mathcal{D}(f)$ in $P\Gamma L(3,q)$.*

Proof. By Theorem 3.4.4, $\mathcal{D}(\psi f) = \bar{\psi}_f\mathcal{D}(f)$ and by hypothesis there exists $\mu \in F^*$ such that $\psi_f f = \mu f$, hence $\mathcal{D}(\mu f) = \bar{\psi}_f\mathcal{D}(f)$, i.e., $\bar{\psi}_f : (1,t,f(t)) \mapsto (1,t,\mu f(t))$. If we compose $\bar{\psi}_f$ with the map sending (x_1,x_2,x_3) to (x_1,x_2,kx_3), with $k = \mu^{-1}$ we see that $\mathcal{D}(f)$ is fixed, hence the theorem follows. □

Corollary 3.4.5 stated in terms of oval covers appears as

Corollary 3.4.6. *Let $f \in \mathcal{F}$ be an o-permutation, and let $\psi \in P\Gamma L(2,q)$. If $\psi\langle f\rangle = \langle f\rangle$, then $\bar{\psi}_{\lambda f}$ is in the stabilizer of $\mathcal{D}(\lambda f)$ in $P\Gamma L(3,q)$ for each $\lambda \in F^*$. Thus, $\bar{\psi}_{\langle f\rangle}$ is in the stabilizer of $\mathcal{D}(\langle f\rangle)$.*

The next theorem is crucial to the further development of the theory. Unfortunately, the proof given in [OP02] is flawed, so we will provide a corrected proof of this result (see [CP03]).

Theorem 3.4.7. *Let f and g be o-permutations for which $\mathcal{D}(f)$ and $\mathcal{D}(g)$ are equivalent under $P\Gamma L(3,q)$. Then there exists $\psi \in P\Gamma L(2,q)$ such that $\psi\langle f\rangle = \langle g\rangle$.*

Proof. For $0 \neq k \in F$, $\mathcal{D}(kf)$ and $\mathcal{D}(g)$ are also equivalent. Suppose $\eta : x \mapsto Bx^\sigma$ ($\in P\Gamma L(3,q)$) satisfies $\eta\mathcal{D}(g) = \mathcal{D}(kf)$. Since η fixes $(0,0,1)$, $\eta(1,0,0) \in \mathcal{D}(kf)$, $\eta(0,1,0) \in \mathcal{D}(kf)$, it follows easily that B has the form

$$\begin{pmatrix} a & b & 0 \\ c & d & 0 \\ akf\left(\frac{c}{a}\right) & bkf\left(\frac{d}{b}\right) & z \end{pmatrix}$$

for some $a, b, c, d, z \in GF(q)$ with $ad + bc \neq 0$ and $z \neq 0$. From

$$\eta\begin{pmatrix} 1 \\ t \\ g(t) \end{pmatrix} = \begin{pmatrix} a + bt^\sigma \\ c + dt^\sigma \\ akf\left(\frac{c}{a}\right) + bkt^\sigma f\left(\frac{d}{b}\right) + zg(t)^\sigma \end{pmatrix}$$

we can compute that

$$g(t)^\sigma = \left(\frac{k}{z}\right)\left[(a + bt^\sigma)f\left(\frac{c + dt^\sigma}{a + bt^\sigma}\right) + bt^\sigma f\left(\frac{d}{b}\right) + af\left(\frac{c}{a}\right)\right].$$

Thus, $g(t)^\sigma = \left(\frac{k}{z}\right)|A'|^{\frac{1}{2}}\psi_0 f(t^\sigma)$ where $\psi_0 \colon x \mapsto A'x$ with $A'\left(\begin{smallmatrix} d & b \\ c & a \end{smallmatrix}\right) \in GL(2,q)$. We then have $g(t) = \left(\frac{k}{z}\right)^{\frac{1}{\sigma}}|A'|^{\frac{1}{2\sigma}}\psi f(t)$ where $\psi \colon x \mapsto (A')^{\frac{1}{\sigma}}x^{\frac{1}{\sigma}}$. Hence, $\left(\frac{z^{\frac{1}{\sigma}}}{|A'|^{\frac{1}{2\sigma}}}\right)g = \psi kf$, and so $\psi\langle f\rangle = \langle g\rangle$. □

Note. It should be noted that the above proof shows that the result is more general than stated. No properties of o-permutations were used. Besides the inclusion of the points $(1,0,0)$ and $(0,1,0)$ in $\mathcal{D}(f)$ and $\mathcal{D}(g)$, the only requirement is that the projectivity between them fixes the point $(0,0,1)$ (whether or not this point is considered to be in the sets or even related to them). This comment also applies to the following corollary.

Corollary 3.4.8. *Let f be an o-permutation in $\mathcal{F}$. Then each element of the stabilizer of $\mathcal{D}(\langle f\rangle)$ is of the form $\bar{\psi}_{\langle f\rangle}$ for some $\psi \in P\Gamma L(2,q)$ such that $\psi\langle f\rangle = \langle f\rangle$.*

Lemma 3.4.9. *Let $\psi \colon x \mapsto Ax^\sigma$ for $A \in GL(2,q)$ and $\sigma \in Aut(F)$. If*

$$\mathcal{C} = \left\{\begin{pmatrix} f(t) & t^{\frac{1}{2}} \\ 0 & g(t) \end{pmatrix} : t \in F\right\} \text{ is a } q-\text{clan},$$

then so is

$$\mathcal{C} = \left\{\begin{pmatrix} \psi f(t) & t^{\frac{1}{2}} \\ 0 & \psi g(t) \end{pmatrix} : t \in F\right\}.$$

Note. This can also be stated in terms of herd covers or herds of ovals. O'Keefe and Penttila normalize the herd cover by insisting that the ovals of the herd always contain the point $(1,1,1)$. So they write their q-clans in the form

$$\mathcal{C} = \left\{ \begin{pmatrix} f_0(t) & t^{\frac{1}{2}} \\ 0 & \kappa f_\infty(t) \end{pmatrix} : t \in F \right\},$$

where $f_0(1) = f_\infty(1) = 1$ and κ is some fixed element with $tr(\kappa) = 1$. In their formulas for $f_0'(t)$, $f_\infty'(t)$ and $f_s'(t)$, they should have written (in their notation and correcting the last two so that $f_\infty'(1) = 1$):

$$\begin{aligned} f_0'(t) &= \psi f_0(t)/\psi f_0(1), \\ f_\infty'(t) &= \psi f_\infty(t)/\psi f_\infty(1), \end{aligned} \tag{3.11}$$

$$f_s'(t) = \frac{f_0'(t) + s^{\frac{1}{2}}t^{\frac{1}{2}} + \kappa' s f_\infty'(t)}{1 + s^{\frac{1}{2}} + \kappa' s}, \text{ where } \kappa' = \psi f_0(1)\psi f_\infty(1)\kappa^\gamma. \tag{3.12}$$

Their lemma then states that if $\{\mathcal{D}(f_s) : s \in \tilde{F}\}$ is a herd of ovals, so is $\{\mathcal{D}(f_s') : s \in \tilde{F}\}$. It then follows that if $\{\mathcal{D}(\langle f_s \rangle) : s \in \tilde{F}\}$ represents a herd cover, so does $\{\mathcal{D}(\langle f_s' \rangle) : s \in \tilde{F}\}$.

3.5 The Automorphism Group of the Herd

We recall here the following definitions:

Definition 3.5.1. *$\mathcal{G}$ is the full group of collineations of $GQ(\mathcal{C})$.*

Definition 3.5.2. *$\mathcal{G}_0$ is the subgroup of $\mathcal{G}$ fixing the points* $(0,0,0)$ *and* (∞).

Definition 3.5.3. *G_0 is the* automorphism group of the herd cover $\mathcal{H}(\mathcal{C})$,*i.e., is the subgroup of $P\Gamma L(2,q)$ that induces a permutation of the oval covers of $\mathcal{H}(\mathcal{C})$.*

Definition 3.5.4. *$\bar{G}_0$ is the* automorphism group of the profile of the herd cover $\overline{\mathcal{H}(\mathcal{C})}$, *i.e., the subgroup of $Aut(G^\otimes)$ that induces a permutation of the ovals of $\overline{\mathcal{H}(\mathcal{C})}$.*

Let $\widehat{\mathcal{H}}(\mathcal{C}) = \{\mathcal{D}(f_s) : s \in \tilde{F}\}$ and $\widehat{\mathcal{H}}(\mathcal{C}') = \{\mathcal{D}(f_t') : t \in \tilde{F}\}$ be herds of two herd covers $\mathcal{H}(\mathcal{C})$ and $\mathcal{H}(\mathcal{C}')$. O'Keefe and Penttila [OP02] define an isomorphism $\theta : \widehat{\mathcal{H}}(\mathcal{C}) \to \widehat{\mathcal{H}}(\mathcal{C}')$ to be a pair $\theta = (\psi, \pi)$ where $\psi \in P\Gamma L(2,q)$ and $\pi : s \mapsto \bar{s}$ is a permutation of the elements of $\tilde{F}$ such that $\psi f_s \in \langle f_{\bar{s}}' \rangle$ for all $s \in \tilde{F}$.

Definition 3.5.5. $\widehat{G_0} = \{\theta : \widehat{\mathcal{H}}(\mathcal{C}) \to \widehat{\mathcal{H}}(\mathcal{C}) : \theta$ *is an isomorphism*$\}$ *is the automorphism group of the herd of ovals (i.e., $Aut(\widehat{\mathcal{H}(\mathcal{C})})$).*

Notes. 1. O'Keefe and Penttila observe that $\widehat{G}_0$ is the stabilizer of $\{\langle f_s \rangle : s \in \tilde{F}\}$ in $P\Gamma L(2,q)$ under the magic action. They go on to calculate the group $\widehat{G}_0$ for the known herds. For both the classical and the FTWKB herds they get the same answer, the group $P\Gamma L(2,q)$. We find this unsatisfactory. In the classical case the group G_0 of the herd cover (which is also the group of the GQ fixing the points (0,0,0) and (∞)) is larger than the group $\widehat{G}_0$ of the herd of ovals. And the induced stabilizer of an oval (i.e., a conic) in the classical case is smaller than the full stabilizer of that conic. This would be impossible to detect under the "magic action" definition since $\widehat{G}_0 = P\Gamma L(2,q)$ in the classical case is the largest possible group permitted by that definition. The classical situation also provides an example of the difference between working with herd covers and working with herds of ovals. In the classical case, the automorphism $\theta = \theta(id, I \otimes P)$ of $G^{\otimes}$ is a collineation of the GQ $\mathcal{S}(\mathcal{C})$ that permutes the ovals $\bar{\mathcal{O}}_s$ according to the rule $s \mapsto s^{-1}$, and so, may be considered as an element of G_0, the automorphism group of the herd cover. The herd cover $\{[\bar{\mathcal{O}}_s] \colon s \in \tilde{F}\}$ of the classical herd has $[\bar{\mathcal{O}}_s] = \{\mathcal{D}(\lambda t^{\frac{1}{2}}) \colon \lambda \in F^*\}$ for each $s \in \tilde{F}$. Now, if we normalize so that $[\bar{\mathcal{O}}_s]$ is represented by $\mathcal{D}(st^{\frac{1}{2}})$ for $s \in F^*$ and by $\mathcal{D}(t^{\frac{1}{2}})$ for $s = 0, \infty$, then θ will be expressed faithfully[1] in the automorphism group of this herd of ovals. On the other hand, if we normalize so that $[\bar{\mathcal{O}}_s]$ is represented by $\mathcal{D}(t^{\frac{1}{2}})$ for all $s \in \tilde{F}$ (which is the normalization used by O'Keefe and Penttila) then θ induces the identity on this herd of ovals and so there is no hope to detect the collineation θ. This cannot happen in a non-classical situation, since then the automorphisms of the herd cover faithfully induce automorphisms of any herd of ovals and $\widehat{G}_0$ and G_0 are the same (see below). Thus, although the two definitions generally lead to the same groups, we believe that our definition of G_0 properly handles the classical case. This is also consistent with the belief that the classical case should be distinguishable from the FTWKB case, where the ovals are non-conical translation ovals.

Let us say some more about the normalization mentioned in Observation 1.6.6 that we often adopt. Since that normalization leads to a faithful action of the automorphism group of the herd of ovals, we believe our choice of normalization is a good one. So, even if we believe that the general setting of the herd cover helps us to understand better the various connections, we also believe that performing such a normalization helps in carrying out easily all the computations without affecting the theory.

2. By the Fundamental Theorem and its extensions, if each element of $\widehat{G}_0$ is really induced by an element of G_0, then the permutation $\pi : s \mapsto \bar{s}$ must be of the form

$$\bar{s} = \frac{as^{\sigma} + b}{cs^{\sigma} + d}$$

for some $\left(\begin{smallmatrix} a & b \\ c & d \end{smallmatrix}\right) \in GL(2,q)$ and some $\sigma \in Aut(F)$. This became obvious once the tensor product representation of the group G_0 had been established.

[1] If G acts on a set B and distinct elements of G induce distinct permutations of B, the action is said to be *faithful*. A faithful action is therefore one in which the associated permutation representation is injective.

3.6 The Groups $\mathcal{G}_0$, G_0 and $\widehat{\mathrm{G}}_0$

The fact that $\mathcal{M}$ is an action implies that $(\mathcal{M}(\psi))^{-1} = \mathcal{M}(\psi^{-1})$. Hence for each nonzero $g \in \mathcal{F}$ there is a nonzero $f \in \mathcal{F}$ for which $\langle g \rangle = \psi\langle f \rangle$.

Recall the notation of Eq. (3.5). Then put $f(t) = \psi^{-1}g(B^T\alpha^\sigma, t)$, i.e., $g(B^T\alpha^\sigma, t) = \psi f$ for $\psi : x \mapsto \left(\begin{smallmatrix} a & b \\ c & d \end{smallmatrix}\right) x^\sigma$. Also, let $\eta = \lambda^2$.

We can then use $\hat{\theta}$ to define

$$\hat{\theta}_{\langle f \rangle}: \begin{pmatrix} 1 \\ t \\ g(\alpha,t) \\ \eta g(\alpha,t) \\ \vdots \\ \eta^{q-2}g(\alpha,t) \end{pmatrix} \mapsto$$

$$\begin{pmatrix} a & b & 0 & 0 & \cdots & 0 \\ c & d & 0 & 0 & \cdots & 0 \\ a\psi f\left(\frac{c}{a}\right) & b\psi f\left(\frac{d}{b}\right) & u & 0 & \cdots & 0 \\ a\psi\eta f\left(\frac{c}{a}\right) & b\psi\eta f\left(\frac{d}{b}\right) & 0 & u & \cdots & 0 \\ \vdots & \vdots & \vdots & \vdots & \ddots & \vdots \\ a\psi\eta^{q-2}f\left(\frac{c}{a}\right) & b\psi\eta^{q-2}f\left(\frac{d}{b}\right) & 0 & 0 & \cdots & u \end{pmatrix} \begin{pmatrix} 1 \\ t \\ g(\alpha,t) \\ \eta g(\alpha,t) \\ \vdots \\ \eta^{q-2}g(\alpha,t) \end{pmatrix}^\sigma . \quad (3.13)$$

Comparing this with Eq. (3.10) we see that

$$\hat{\theta}_{\langle f \rangle} = \bar{\psi}_{\langle f \rangle}. \quad (3.14)$$

By Theorem 3.4.4 f is an o-permutation and $\bar{\psi}_{\langle f \rangle} \colon \mathcal{D}_\eta(f) \mapsto \mathcal{D}_{\eta^\sigma}(\psi f) = \mathcal{D}_{\eta^\sigma}(g(B^T\alpha^\sigma, t))$. Since

$$\bar{\psi}_{\langle f \rangle} = \hat{\theta}_{\langle f \rangle} : \mathcal{D}_\eta(g(\alpha,t)) \mapsto \mathcal{D}_{\eta^\sigma}(g(B^T\alpha^\sigma, t)), \quad (3.15)$$

it must be that $\mathcal{D}_\eta(f) = \mathcal{D}_\eta(g(\alpha,t))$. It follows that $\langle f(t) \rangle = \langle g(\alpha,t) \rangle$.

This proves the following important theorem.

Theorem 3.6.1. *If $\mathcal{C}$ is a q-clan, then*

$$\theta\left(\sigma, (A^T)^{(\frac{1}{2})} \otimes (B^T)^{(\frac{1}{2})}\right) \in \mathcal{G} \text{ iff } \psi\langle g(\alpha,t)\rangle = \langle g(B\alpha^\sigma, t)\rangle$$

for all $\alpha \in PG(1,q)$, where $\psi \colon x \mapsto Ax^\sigma$. □

In [OP02] $\pi(\mathcal{C})$ is defined to be the subspace of $\mathcal{F}$ spanned by $f(t)$, $t^{\frac{1}{2}}$ and $g(t)$, the q-clan functions. In the classical case $\pi(\mathcal{C})$ is 1-dimensional and for any $A \in GL(2,q)$, $\sigma \in Aut(F)$ and any $B \in SL(2,q)$, if $\psi : x \mapsto (A^{(2)})^T x^\sigma$, automatically $\psi\langle g(\alpha,t)\rangle = \langle g((B^{(2)})^T\alpha^\sigma, t)\rangle$. However, following Lemma 13 of [OP02] it is shown that if $\mathcal{C}$ is non-classical, then $\pi(\mathcal{C})$ is 3-dimensional and no two herd functions are even in the same 1-dimensional space. Hence if ψ is given, and σ is given, clearly only one B can exist. (This also follows from Theorem 1.10.3 of [Pa98].) This then implies

Corollary 3.6.2. *If $\mathcal{C}$ is a non-classical q-clan, then $G_0 = \widehat{G}_0$.*

3.7 The Square-Bracket Function

As always throughout these notes, $q = 2^e$. Let $F = GF(q) \subseteq GF(q^2) = E$. Write $\overline{x} = x^q$ for $x \in E$. So $x = \overline{x}$ iff $x \in F$. Let ζ be a primitive element of E, so the multiplicative order $|\zeta|$ of ζ is $|\zeta| = q^2 - 1$. Put $\beta = \zeta^{q-1}$, so $|\beta| = q + 1$. Then $\bar{\beta} = \beta^q = \beta^{-1}$. Define $\delta = \beta + \bar{\beta}$. More generally, for each rational number a with denominator relatively prime to $q + 1$ we use the following convenient notation:

$$[a] := \beta^a + \bar{\beta}^a. \tag{3.16}$$

Lemma 3.7.1. *The following are easy consequences of the definition:*

(i) $\delta = [1]$; $0 = [0]$.

(ii) $[a] = [b]$ *iff* $a \equiv \pm b \pmod{q + 1}$.

(iii) $[a] \cdot [b] = [a + b] + [a - b]$; $[a] + [b] = \left[\frac{a+b}{2}\right] \cdot \left[\frac{a-b}{2}\right]$.

(iv) $[a]^\sigma = [\sigma a]$ *for* $\sigma = 2^i \in Aut(F)$.

(v) $\left[\frac{a+c}{2}\right]\left[\frac{a}{2}\right]\left[\frac{c}{2}\right] = [a + c] + [a] + [c]$.

(vi) $[3a] = [a]^3 + [a]$; $[5a] = [a]^5 + [a]^3 + [a]$.

(vii) *The map* $\frac{[j+k]}{[j]} \mapsto \frac{[j+1+k]}{[j+1]}$, *for all* $j \in \tilde{F}$, *permutes the elements of* $\tilde{F}$ *in a cycle of length* $q + 1$. *If* $k = 1$, *the map* $\frac{[j+1]}{[j]} \mapsto \frac{[j+2]}{[j+1]}$, *for all* $j \in \tilde{F}$, *is the same as the map* $t \mapsto t^{-1} + \delta$. *More generally,* $\frac{[j+k]}{[j]} \mapsto \frac{[j+1+k]}{[j+1]}$ *is the map* $t \mapsto \frac{t[k+1]+[1]}{t[1]+[k-1]}$.

It will also be useful to have a generalization of the above notation. For each k, $1 \leq k \leq q$, put $\beta_k = \zeta^{k(q-1)}$, so $|\beta_k| = \frac{q+1}{gcd(k,\ q+1)}$. Then $\overline{\beta_k} = {\beta_k}^q = {\beta_k}^{-1}$. Let $\delta_k = \beta_k + \overline{\beta_k}$ and note that $[ka] = \beta_k^a + \overline{\beta_k^a}$. When $k = 1$ we suppress the subscript. Here we are assuming that $k = 1$. Since $\beta^j \in F$ if and only if $j \equiv 0 \pmod{q + 1}$, for $1 \leq j \leq q$, $x^2 + [j]x + 1 = (x + \beta^j)(x + \bar{\beta}^j)$ is irreducible over F. Hence we have

$$tr\left(\frac{1}{[j]}\right) = tr\left(\frac{1}{[j]^2}\right) = tr\left(\frac{1}{[2j]}\right) = 1. \tag{3.17}$$

Note. Since $[a] = [b]$ iff $a \equiv \pm b \pmod{q + 1}$ it follows that the elements of F with trace equal to 1 are precisely those of the form $\frac{1}{[j]}$, $1 \leq j \leq q$.

Put $M = \left(\begin{smallmatrix} 0 & 1 \\ 1 & \sqrt{\delta} \end{smallmatrix}\right)$ with eigenvalues $\beta^{\pm\frac{1}{2}}$. As M^j has eigenvalues $\beta^{\pm\frac{j}{2}}$, which belong to F iff $j \equiv 0 \pmod{q+1}$ iff $\beta^{\pm\frac{j}{2}} = 1$, we see that $M^{q+1} = I$, and in fact M has multiplicative order $q+1$. Note also that M is symmetric with $\det(M) = 1$. Also M has left eigenvectors $(1, \beta^{\frac{1}{2}})$ and $(\beta^{\frac{1}{2}}, 1)$ belonging, respectively, to eigenvalues $\beta^{\frac{1}{2}}$ and $\beta^{-\frac{1}{2}}$. It follows that

$$M = \frac{1}{1+\beta}\begin{pmatrix} 1 & \beta^{\frac{1}{2}} \\ \beta^{\frac{1}{2}} & 1 \end{pmatrix}\begin{pmatrix} \beta^{\frac{1}{2}} & 0 \\ 0 & \beta^{-\frac{1}{2}} \end{pmatrix}\begin{pmatrix} 1 & \beta^{\frac{1}{2}} \\ \beta^{\frac{1}{2}} & 1 \end{pmatrix}. \tag{3.18}$$

For any integer j we use Eq. (3.18) to compute

$$M^j = \frac{1}{1+\beta}\begin{pmatrix} 1 & \beta^{\frac{1}{2}} \\ \beta^{\frac{1}{2}} & 1 \end{pmatrix}\begin{pmatrix} \beta^{\frac{j}{2}} & 0 \\ 0 & \beta^{\frac{-j}{2}} \end{pmatrix}\begin{pmatrix} 1 & \beta^{\frac{1}{2}} \\ \beta^{\frac{1}{2}} & 1 \end{pmatrix} = \frac{1}{\delta^{\frac{1}{2}}}\begin{pmatrix} [\frac{j-1}{2}] & [\frac{j}{2}] \\ [\frac{j}{2}] & [\frac{j+1}{2}] \end{pmatrix}. \tag{3.19}$$

With $P = \left(\begin{smallmatrix} 0 & 1 \\ 1 & 0 \end{smallmatrix}\right)$,

$$M^j P M^{-j} = \frac{1}{\delta^{\frac{1}{2}}}\begin{pmatrix} [j] & [j-\frac{1}{2}] \\ [j+\frac{1}{2}] & [j] \end{pmatrix}, \tag{3.20}$$

and it has unique eigenvector $([\frac{2j+1}{4}], [\frac{2j-1}{4}])$.

Our interest in the matrix M derives from the fact that since it has multiplicative order $q+1$, for some integer m (modulo $q+1$) there could be a collineation of $GQ(\mathcal{C})$ of the form $\theta(id, M \otimes M^{-m})$ permuting the lines through (∞) in a cycle of length $q+1$. We explore this possibility in the next section.

Occasionally it is of interest to know which is the smallest subfield of $E = GF(q^2)$ that contains δ_k.

Exercise 3.7.2. *Let ζ be a primitive element for $E = GF(q^2) \supseteq GF(q) = F$, and let $1 \le k \le q$. Put $\beta_k = \zeta^{k(q-1)}$ and $\delta_k = \beta_k + \bar{\beta}_k$. Then if $r = gcd(k, q+1)$, the field $GF(2)(\delta_k)$ is the field $GF(2^d)$ where d is the smallest divisor of e for which $\frac{e}{d}$ is odd and $\frac{2^e+1}{2^d+1} | r$. In particular, if k and $q+1$ are relatively prime, then $GF(2)(\delta_k) = GF(q)$.*

3.8 A Cyclic Linear Collineation

Let m be a positive integer (modulo $q+1$). If there is a collineation of $GQ(\mathcal{C})$ of the form $\theta = \theta(id, M \otimes M^{-m})$, then by Theorem 2.3.4, θ has the following effect:

$$\theta = \theta(id, M \otimes M^{-m}) : ((\alpha, \beta), c) \mapsto$$
$$((\alpha, \beta)(M \otimes M^{-m}), c + \alpha \circ \beta + \beta M^{-m} A_\delta M^{-m} \beta^T). \tag{3.21}$$

A little work shows that if $t = \frac{[k+1]}{[k]}$ (replacing $\frac{\alpha}{[k]^{1/2}}$ by α), then the typical element of $A(t)$ can be represented in the form

$$\left(\left(\left[\frac{k}{2}\right], \left[\frac{k+1}{2}\right]\right) \otimes \alpha, [k]g(\alpha, t)\right), \quad \alpha \in F^2.$$

It follows that

$$\theta = \theta(id, M \otimes M^{-m}) =: \left(\left(\left[\frac{k}{2}\right], \left[\frac{k+1}{2}\right]\right) \otimes \alpha, [k]g(\alpha, t)\right) \mapsto$$
$$\left(\left(\left[\frac{k}{2}\right], \left[\frac{k+1}{2}\right]\right) M \otimes \alpha M^{-m}, [k]g(\alpha, t) + [k+1]\alpha M^{-m} A_\delta M^{-m} \alpha^T\right)$$
$$= \left(\left(\left[\frac{k+1}{2}\right], \left[\frac{k+2}{2}\right]\right) \otimes \alpha M^{-m}, [k]g(\alpha, t) + [k+1]\alpha M^{-m} A_\delta M^{-m} \alpha^T\right).$$

So what does it mean for this to be in $A(\bar{t})$ where $\bar{t}t^{-1} + \delta$? Since $t = \frac{[k+1]}{[k]}$, $\bar{t} = \frac{[k+2]}{[k+1]}$. So containment holds iff

$$\left(\left(\left[\frac{k+1}{2}\right], \left[\frac{k+2}{2}\right]\right) \otimes \alpha M^{-m}, [k]g(\alpha, t) + [k+1]\alpha M^{-m} A_\delta M^{-m} \alpha^T\right)$$
$$= \left(\left(\left[\frac{k+1}{2}\right], \left[\frac{k+2}{2}\right]\right) \otimes \alpha M^{-m}, [k+1]g(\alpha M^{-m}, t^{-1} + \delta)\right)$$

iff

$$\alpha \left\{[k+1]M^{-m} A_{t^{-1}+\delta} M^{-m}\right\} \alpha^T = \alpha \left\{[k]A_t + [k+1]M^{-m} A_\delta M^{-m}\right\} \alpha^T$$

iff

$$\frac{[k+1]}{[k]} A_{t^{-1}+\delta} \equiv M^m A_t M^m + \frac{[k+1]}{[k]} A_\delta \quad \text{for all } t \in F$$

iff

$$M^m A_t M^m \equiv t\,(A_{t^{-1}+\delta} + A_\delta)\,.$$

A routine computation shows that $M^m A_t M^m \equiv$

$$\frac{1}{\delta}\begin{pmatrix} [m-1]f(t) + \left[\frac{m-1}{2}\right]\left[\frac{m}{2}\right] t^{\frac{1}{2}} + [m]g(t) & \delta t^{\frac{1}{2}} \\ 0 & [m]f(t) + \left[\frac{m}{2}\right]\left[\frac{m+1}{2}\right] t^{\frac{1}{2}} + [m+1]g(t) \end{pmatrix}. \tag{3.22}$$

This proves the following theorem.

Theorem 3.8.1. *The automorphism $\theta(id, M \otimes M^{-m})$ of $G^{\otimes}$ induces a collineation of $GQ(\mathcal{C})$ if and only if:*

$$\text{(i) } t = \left\{f(t^{-1} + \delta) + f(\delta)\right\} = \frac{[m-1]}{\delta}\left\{f(t) + \left(\frac{[m]}{[m-1]}\right)^{\frac{1}{2}} t^{\frac{1}{2}} + \left(\frac{[m]}{[m-1]}\right) g(t)\right\}$$

and

$$\text{(ii) } t = \left\{g(t^{-1} + \delta) + g(\delta)\right\} = \frac{[m]}{\delta}\left\{f(t) + \left(\frac{[m+1]}{[m]}\right)^{\frac{1}{2}} t^{\frac{1}{2}} + \frac{[m+1]}{[m]} g(t)\right\}.$$

Theorem 3.8.2. *Suppose that $\theta = \theta(id, M \otimes M^{-m})$ is a collineation of $GQ(\mathcal{C})$. Then θ has order $q+1$ on the lines of $GQ(\mathcal{C})$ through the point (∞). For each integer j modulo $q+1$ we have the following:*

(i) $\theta^j = \theta(id, M^j \otimes M^{-mj}) : [A(t)] \mapsto [A(\bar{t})]$, *where* $\bar{t} = \frac{[j+1]t+[j]}{[j]t+[j-1]}$.

If $t = \frac{[k+1]}{[k]}$, *then* $\bar{t} = \frac{[j+k+1]}{[j+k]} = \frac{(1+\delta s)t+s}{st+1}$ *if* $s = \frac{[j]}{[j-1]}$.

(ii) $\theta^j : \mathcal{O}_t \mapsto \mathcal{O}_{\bar{t}}$ *where* $\bar{t} = \frac{[mj-1]t+[mj]}{[mj]t+[mj+1]}$.

If $t = \frac{[k+1]}{[k]}$, *then* $\bar{t}\frac{[k-mj+1]}{[k-mj]} = \frac{(1+\delta r)t+r}{rt+1}$, *if* $r = \frac{[mj]}{[mj+1]}$.

3.9 Some Involutions

From now on we will assume that the automorphisms of $G^{\otimes}$ given by $\theta = \theta(id, M \otimes M^{-m})$ and $\varphi = \theta(id, P \otimes P)$ really do induce collineations of $GQ(\mathcal{C})$. Recall that φ is an involution interchanging $[A(t)]$ and $[A(t^{-1})]$ and interchanging $\mathcal{O}_s$ and $\mathcal{O}_{s^{-1}}$, for $t, s \in \tilde{F}$.

Theorem 3.9.1. *Put*

$$\theta_j = \theta^j \circ \varphi \circ \theta^{-j} = \theta(id, M^j P M^{-j} \otimes M^{-mj} P M^{mj}).$$

Then θ_j is an involution fixing

$$[A(s)] \textit{ with } s = \frac{[j-\frac{1}{2}]}{[j+\frac{1}{2}]},$$

and fixing

$$O_r \textit{ with } r = \frac{[mj+\frac{1}{2}]}{[mj-\frac{1}{2}]}.$$

In general $\theta_j : [A(t)] \mapsto [A(\bar{t})]$ with

$$\bar{t} = \frac{[2j]t+[2j-1]}{[2j+1]t+[2j]} = \frac{t(s^2+1)+\delta s^2}{t\delta+s^2+1}.$$

Similarly, $\theta_j : O_t \mapsto O_{\bar{t}}$ with

$$\bar{t} = \frac{[2mj]t+[2mj+1]}{[2mj-1]t+[2mj]} = \frac{t(r^2+1)+\delta r^2}{t\delta+r^2+1}.$$

If we use the notation that I_s is the involution above fixing the line $[A(s)]$, then the following lemma is sometimes useful.

Lemma 3.9.2. *For* $s = \frac{[j-\frac{1}{2}]}{[j+\frac{1}{2}]}$,

$$I_s = \theta_j = \theta\left(id, \frac{1}{\delta^{\frac{1}{2}}}\begin{pmatrix} [j] & [j-\frac{1}{2}] \\ [j+\frac{1}{2}] & [j] \end{pmatrix} \otimes \frac{1}{\delta^{\frac{1}{2}}}\begin{pmatrix} [mj] & [mj+\frac{1}{2}] \\ [mj-\frac{1}{2}] & [mj] \end{pmatrix}\right).$$

Moreover, using $j = 0$ *and* $j = -\frac{1}{2}$ *we find* $I_1 \circ I_\infty = \theta(id, M \otimes M^{-m})$.

Proof. Simple substitution in the preceding formulas and a short computation. □

The following observations about involutions I_s are sometimes useful.

Lemma 3.9.3. *Suppose that* $\theta = \theta(id, A \otimes B)$ *is the involution* I_s *with*

$$A = \begin{pmatrix} a_4 & a_2 \\ a_3 & a_1 \end{pmatrix} \text{ and } B = \begin{pmatrix} b_4 & b_2 \\ b_3 & b_1 \end{pmatrix}.$$

Then $\theta : A(t) \mapsto A(\bar{t})$ *where* $(1, \bar{t}) \equiv (1, t)A^{(2)}$ *implies* $\bar{t} = \frac{a_1^2 t + a_2^2}{a_3^2 t + a_4^2} = \frac{t(s^2+1)+\delta s^2}{t\delta + s^2 + 1}$. *Also,* $A(s)$ *is fixed with* $s = \frac{a_2}{a_3}$, $a_4 = a_1$, *and* $a_1\delta^{\frac{1}{2}} + a_2 + a_3 = 0$.

Similarly, $\mathcal{O}_t \mapsto \mathcal{O}_{\bar{t}}$ *where* $(1, \bar{t}) \equiv (1, t)B^{(2)}$ *implies* $\bar{t} = \frac{b_1^2 t + b_2^2}{b_3^2 t + b_4^2} = \frac{t(r^2+1)+\delta r^2}{t\delta + r^2 + 1}$. *Here* $\mathcal{O}_r$ *is fixed with* $r = \frac{b_2}{b_3}$, $b_4 = b_1$, *and* $b_1\delta^{\frac{1}{2}} + b_2 + b_3 = 0$.

Proof. We will show that $A(s)$ is fixed with $s = \frac{a_2}{a_3}$, $a_4 = a_1$, and $a_1\delta^{\frac{1}{2}} + a_2 + a_3 = 0$. By Theorem 3.9.1 and the note following Theorem 2.3.4, we know that $\frac{a_1^2 t + a_2^2}{a_3^2 t + a_4^2} = \frac{t(s^2+1)+\delta s^2}{t\delta + s^2 + 1}$, which is equivalent to $\frac{\left(\frac{a_1}{a_3}\right)^2 t + \left(\frac{a_2}{a_3}\right)^2}{t + \left(\frac{a_4}{a_3}\right)^2} = \frac{\left(\frac{s^2+1}{\delta}\right)t + s^2}{t + \left(\frac{s^2+1}{\delta}\right)}$. So it follows that $\left(\frac{a_2}{a_3}\right)^2 = s^2$, $\left(\frac{a_1}{a_3}\right)^2 = \frac{s^2+1}{\delta}$ and $\left(\frac{a_4}{a_3}\right)^2 = \frac{s^2+1}{\delta}$ which give $a_4 = a_1$ and $a_1\delta^{\frac{1}{2}} + a_2 + a_3 = 0$. Similarly, we can prove the second part of the theorem. □

(As a check, the last equality applied to θ_j becomes $[mj]\left[\frac{1}{2}\right] = \left[mj + \frac{1}{2}\right] + \left[mj - \frac{1}{2}\right]$, which is clearly true.)

3.10 Some Semi-linear Collineations

Assume that $\mathcal{C}$ is a given q-clan for which the conditions in Theorems 3.2.1 and 3.8.1 both hold. In the non-classical case it can happen that each line through (∞) (resp., each oval $\mathcal{O}_t$) is fixed by a *unique* involution. Even when this is not true, however, it is profitable to assume that an involution that turns up which fixes $[A(\infty)]$ is I_∞, the one we know about. In that case we ask for necessary and sufficient conditions on the functions f and g that give $\mathcal{C}$ so that there is a collineation of $GQ(\mathcal{C})$ that fixes the line $[A(\infty)]$ and the points (∞) and $(0,0,0)$ and belongs to the field automorphism σ. From Eq. (2.42) we see that such a collineation must be of the form (we use the symbol $\hat{\theta}$ hoping that the context makes clear when it is referred to an induced collineation or to a semi-linear collineation)

$$\hat{\theta} = \theta(\sigma, \begin{pmatrix} 1 & \overline{0}^{\frac{1}{2}} \\ 0 & \lambda \end{pmatrix} \otimes B), \text{ where}$$

$$\hat{\theta} : A(t) \mapsto A(\overline{t}) \text{ with } \overline{t} = \lambda^2 t^{\sigma} + \overline{0}, \text{ and } \det(B) = 1.$$

Put $B = \begin{pmatrix} a & b \\ c & d \end{pmatrix}$. The involution we know about that fixes $[A(\infty)]$ is

$$\theta_{\frac{-1}{2}} = I_{\infty} = \theta\left(id, \begin{pmatrix} 1 & \delta^{\frac{1}{2}} \\ 0 & 1 \end{pmatrix} \otimes \frac{1}{\delta^{\frac{1}{2}}} \begin{pmatrix} \left[\frac{m}{2}\right] & \left[\frac{m-1}{2}\right] \\ \left[\frac{m+1}{2}\right] & \left[\frac{m}{2}\right] \end{pmatrix}\right).$$

Since I_{∞} also fixes the unique oval O_r with $r = \left[\frac{m-1}{2}\right]/\left[\frac{m+1}{2}\right]$, it follows that any collineation that fixes $[A(\infty)]$ must also fix this same oval.

Straightforward but somewhat tiresome calculations establish the following:

Lemma 3.10.1. (i) $\hat{\theta}^{-1} = \theta\left(\sigma^{-1}, \begin{pmatrix} 1 & \overline{0}^{\frac{1}{2\sigma}} \lambda^{-\frac{1}{\sigma}} \\ 0 & \lambda^{-\frac{1}{\sigma}} \end{pmatrix} \otimes \begin{pmatrix} d^{\frac{1}{\sigma}} & b^{\frac{1}{\sigma}} \\ c^{\frac{1}{\sigma}} & a^{\frac{1}{\sigma}} \end{pmatrix}\right)$.

(ii) $\hat{\theta} \circ \theta_{-\frac{1}{2}} \circ \hat{\theta}^{-1} \theta\left(id, \begin{pmatrix} 1 & \delta^{\frac{1}{2\sigma}} \lambda^{-\frac{1}{\sigma}} \\ 0 & 1 \end{pmatrix} \otimes\right.$

$$\left.\frac{1}{\delta^{\frac{1}{2\sigma}}} \begin{pmatrix} \left[\frac{m}{2\sigma}\right] + (ac)^{\frac{1}{\sigma}}\left[\frac{m-1}{2\sigma}\right] + (bd)^{\frac{1}{\sigma}}\left[\frac{m+1}{2\sigma}\right] & a^{\frac{2}{\sigma}}\left[\frac{m-1}{2\sigma}\right] + b^{\frac{2}{\sigma}}\left[\frac{m+1}{2\sigma}\right] \\ c^{\frac{2}{\sigma}}\left[\frac{m-1}{2\sigma}\right] + d^{\frac{2}{\sigma}}\left[\frac{m+1}{2\sigma}\right] & \left[\frac{m}{2\sigma}\right] + (ac)^{\frac{1}{\sigma}}\left[\frac{m-1}{2\sigma}\right] + (bd)^{\frac{1}{\sigma}}\left[\frac{m+1}{2\sigma}\right] \end{pmatrix}\right).$$

It is clear that $\hat{\theta} \circ \theta_{-\frac{1}{2}} \circ \hat{\theta}^{-1}$ must be an involution fixing $[A(\infty)]$, hence we assume that it equals I_{∞} given above. In the presentation of the collineation, both factors in the tensor product have determinant 1 so are uniquely determined. Hence by comparing the entries in the (1,2) positions of the left factors, we obtain that $\delta^{\frac{1}{2}} = \delta^{\frac{1}{2\sigma}} \lambda^{-\frac{1}{\sigma}}$, from which we see

$$\lambda = \delta^{\frac{1-\sigma}{2}}. \tag{3.23}$$

First we consider the result of the equality of this involution with I_{∞}. By equating right-hand factors in the tensor products we obtain the following:

$$\left[\frac{m}{2}\right] = \delta^{\frac{\sigma-1}{2\sigma}}\left[\frac{m}{2\sigma}\right] + \delta^{\frac{\sigma-1}{2\sigma}}(ac)^{\frac{1}{\sigma}}\left[\frac{m-1}{2\sigma}\right] + \delta^{\frac{\sigma-1}{2\sigma}}(bd)^{\frac{1}{\sigma}}\left[\frac{m+1}{2\sigma}\right]. \tag{3.24}$$

$$\left[\frac{m-1}{2}\right] = \delta^{\frac{\sigma-1}{2\sigma}} a^{\frac{2}{\sigma}}\left[\frac{m-1}{2\sigma}\right] + \delta^{\frac{\sigma-1}{2\sigma}} b^{\frac{2}{\sigma}}\left[\frac{m+1}{2\sigma}\right]. \tag{3.25}$$

$$\left[\frac{m+1}{2}\right] = \delta^{\frac{\sigma-1}{2\sigma}} c^{\frac{2}{\sigma}}\left[\frac{m-1}{2\sigma}\right] + \delta^{\frac{\sigma-1}{2\sigma}} d^{\frac{2}{\sigma}}\left[\frac{m+1}{2\sigma}\right]. \tag{3.26}$$

$$ad + bc = 1. \tag{3.27}$$

It is straightforward to show that Eq. (3.27) applied to the product of Eqs. (3.25) and (3.26) implies Eq. (3.24). Also, we can use these equations to solve for b and d as linear functions of a and c:

$$b = \frac{\delta^{\frac{1-\sigma}{4}} \left[\frac{m-1}{2}\right]^{\frac{q}{2}} + a \left[\frac{m-1}{2\sigma}\right]^{\frac{q}{2}}}{\left[\frac{m+1}{4}\right]}. \tag{3.28}$$

$$d = \frac{\delta^{\frac{1-\sigma}{4}} \left[\frac{m+1}{2}\right]^{\frac{q}{2}} + c \left[\frac{m-1}{2\sigma}\right]^{\frac{q}{2}}}{\left[\frac{m+1}{4}\right]}. \tag{3.29}$$

Put these expressions for b and d in $1 = ad + bc$ and solve for c to obtain

$$c = \frac{\delta^{\frac{\sigma-1}{4}} \left[\frac{m+1}{4}\right] + a \left[\frac{m+1}{2}\right]^{\frac{q}{2}}}{\left[\frac{m-1}{2}\right]^{\frac{q}{2}}}. \tag{3.30}$$

Now use Eqs. (3.29) and (3.30) to express d as a linear function of a:

$$d = \delta^{\frac{1-\sigma}{4}} \frac{\left[\frac{m}{2}\right]^{\sigma-1} + a \left[\frac{m+1}{4}\right]^{\sigma-1}}{\left[\frac{m-1}{4}\right]^{\sigma-1}}. \tag{3.31}$$

At this point we have the entries of the matrix B written as linear functions of the entry a.

Lemma 3.10.2. (i) *For arbitrary j modulo $q+1$ we have*

$$\begin{aligned}
&\hat{\theta}^{-1} \circ \theta_j \circ \hat{\theta} \\
&= \theta\left(id, \frac{1}{\delta^{\frac{q}{2}}} \begin{pmatrix} [j]^\sigma + \bar{0}^{\frac{1}{2}}\lambda^{-1}[j+\frac{1}{2}]^\sigma & \bar{0}\lambda^{-1}[j+\frac{1}{2}]^\sigma + \lambda[j-\frac{1}{2}]^\sigma \\ \lambda^{-1}[j+\frac{1}{2}]^\sigma & [j]^\sigma + \bar{0}^{\frac{1}{2}}\lambda^{-1}[j+\frac{1}{2}]^\sigma \end{pmatrix}\right. \\
&\left.\otimes \frac{1}{\delta^{\frac{q}{2}}} \begin{pmatrix} [mj]^\sigma + ab[mj-\frac{1}{2}]^\sigma + cd[mj+\frac{1}{2}]^\sigma & b^2[mj-\frac{1}{2}]^\sigma + d^2[mj+\frac{1}{2}]^\sigma \\ a^2[mj-\frac{1}{2}]^\sigma + c^2[mj+\frac{1}{2}]^\sigma & [mj]^\sigma + ab[mj-\frac{1}{2}]^\sigma + cd[mj+\frac{1}{2}]^\sigma \end{pmatrix}\right)
\end{aligned}$$

where $\lambda = \delta^{\frac{1-\sigma}{2}}$.

(ii) $\hat{\theta}^{-1} \circ \theta_j \circ \hat{\theta} : A(t) \mapsto A(\bar{t})$ *where*

$$\begin{aligned}
t \overset{\hat{\theta}^{-1}}{\mapsto} ((t+\bar{0})\delta^{\sigma-1})^{\frac{1}{\sigma}} &\overset{\theta_j}{\mapsto} \frac{((t+\bar{0})\delta^{\sigma-1})^{\frac{1}{\sigma}}(s^2+1)+\delta s^2}{((t+\bar{0})\delta^{\sigma-1})^{\frac{1}{\sigma}}\delta+s^2+1} \\
&\overset{\hat{\theta}}{\mapsto} \delta^{1-\sigma}\left\{\frac{(t+\bar{0})(s^2+1)^\sigma\delta^{\sigma-1}+\delta^\sigma s^{2\sigma}}{(t+\bar{0})\delta^\sigma.\delta^{\sigma-1}+s^{2\sigma}+1}\right\} + \bar{0} \\
&= \frac{t(\bar{0}\delta+\delta^{2-2\sigma}(s^{2\sigma}+1))+\delta^{3-2\sigma}s^{2\sigma}+\bar{0}^2\delta}{t\delta+\bar{0}\delta+\delta^{2-2\sigma}(s^{2\sigma}+1)}, \text{ where } s = \frac{[j-\frac{1}{2}]}{[j+\frac{1}{2}]}.
\end{aligned}$$

This expression must be of the form

$$\frac{t(\bar{s}^2+1)+\delta\bar{s}^2}{t\delta+\bar{s}^2+1}.$$

It follows that $\delta\overline{s}^2 = \delta^{3-2\sigma}s^{2\sigma} + \overline{0}^2\delta$ *and* $\overline{s}^2 + 1 = \overline{0}\delta + \delta^{2-2\sigma}(s^{2\sigma}+1)$. *From this it follows that*

$$\overline{0}^2 + \overline{0}\delta + \delta^{2-2\sigma} + 1 = 0. \tag{3.32}$$

Equation (3.32) *has two solutions:*

$$\overline{0} = \sum_{i=1}^{j} [1]^{2^i} = \frac{[\sigma-1]}{[\sigma]} \quad \textit{where } \sigma \colon x \mapsto x^{2^j}, \tag{3.33}$$

and

$$\overline{0} = \frac{[\sigma-1]}{[\sigma]} + [1]. \tag{3.34}$$

Note. $\frac{[\sigma-1]}{[\sigma]} = \frac{[(\sigma-\frac{1}{2})-\frac{1}{2}]}{[(\sigma-\frac{1}{2})+\frac{1}{2}]}$ and $\frac{[\sigma-1]}{[\sigma]} + [1] = \frac{[\sigma+1]}{[\sigma]} = \left(\frac{[(\sigma+\frac{1}{2})-\frac{1}{2}]}{[(\sigma+\frac{1}{2})+\frac{1}{2}]}\right)^{-1}$.

Put $j = \frac{1}{2}$, so $s = 0$. Then $\hat{\theta}^{-1} \circ \theta_{\frac{1}{2}} \circ \hat{\theta}$ is the involution fixing $A(\overline{0})$.

We will be concerned with the case $\overline{0} = \frac{[\sigma-1]}{[\sigma]} = \frac{[(\sigma-\frac{1}{2})-\frac{1}{2}]}{[(\sigma-\frac{1}{2})+\frac{1}{2}]}$. Hence the involution $\hat{\theta}^{-1} \circ \theta_{\frac{1}{2}} \circ \hat{\theta}$ must be $\theta_{\sigma-\frac{1}{2}}$. So $\hat{\theta}^{-1} \circ \theta_{\frac{1}{2}} \circ \hat{\theta} = \theta_{\sigma-\frac{1}{2}}$, which in fact says

$$\begin{aligned}
&\theta\left(id, \frac{1}{\delta^{\frac{\sigma}{2}}}\begin{pmatrix} [\frac{1}{2}]^\sigma + \frac{[\sigma-1]^{1/2}}{[\sigma]^{1/2}}[1]^{(3\sigma-1)/2} & [1]^{\frac{3\sigma-1}{2}}\frac{[\sigma-1]}{[\sigma]} \\ [1]^{\frac{3\sigma-1}{2}} & [1]^{(3\sigma-1)/2}\frac{[\sigma-1]^{1/2}}{[\sigma]^{1/2}} + [\frac{1}{2}]^\sigma \end{pmatrix}\right. \\
&\quad\left.\otimes \frac{1}{\delta^{\frac{\sigma}{2}}}\begin{pmatrix} [\frac{m}{2}]^\sigma + ab[\frac{m-1}{2}]^\sigma + cd[\frac{m+1}{2}]^\sigma & b^2[\frac{m-1}{2}]^\sigma + d^2[\frac{m+1}{2}]^\sigma \\ a^2[\frac{m-1}{2}]^\sigma + c^2[\frac{m+1}{2}]^\sigma & [\frac{m}{2}]^\sigma + ab[\frac{m-1}{2}]^\sigma + cd[\frac{m+1}{2}]^\sigma \end{pmatrix}\right) \\
&= \theta\left(id, \frac{1}{\delta^{\frac{1}{2}}}\begin{pmatrix} [\sigma-\frac{1}{2}] & [\sigma-1] \\ [\sigma] & [\sigma-\frac{1}{2}] \end{pmatrix} \otimes \frac{1}{\delta^{\frac{1}{2}}}\begin{pmatrix} [m(\sigma-\frac{1}{2})] & [m(\sigma-\frac{1}{2})+\frac{1}{2}] \\ [m(\sigma-\frac{1}{2})-\frac{1}{2}] & [m(\sigma-\frac{1}{2})] \end{pmatrix}\right).
\end{aligned}$$

Equating the two right-hand factors in the tensor products yields the following three equations:

$$[1]^{\frac{\sigma-1}{2}}\left[m(\sigma-\frac{1}{2})\right] = \left[\frac{m}{2}\right]^\sigma + ab\left[\frac{m-1}{2}\right]^\sigma + cd\left[\frac{m+1}{2}\right]^\sigma. \tag{3.35}$$

$$[1]^{\frac{\sigma-1}{2}}\left[m(\sigma-\frac{1}{2})+\frac{1}{2}\right] = b^2\left[\frac{m-1}{2}\right]^\sigma + d^2\left[\frac{m+1}{2}\right]^\sigma. \tag{3.36}$$

$$[1]^{\frac{\sigma-1}{2}}\left[m(\sigma-\frac{1}{2})-\frac{1}{2}\right] = a^2\left[\frac{m-1}{2}\right]^\sigma + c^2\left[\frac{m+1}{2}\right]^\sigma. \tag{3.37}$$

Now take Equations 3.30 and 3.37, rearrange them slightly, square the latter and raise the former to the fourth power to obtain the following two equations:

$$a^4[m+1]^\sigma + c^4[m-1]^\sigma = [m+1][1]^{\sigma-1}. \tag{3.38}$$

$$a^4[m-1]^\sigma + c^4[m+1]^\sigma = [m(2\sigma-1)-1][1]^{\sigma-1}. \tag{3.39}$$

Solving these two linear equations in two unknowns we obtain

$$a^4 = \frac{[m(\sigma-1)-(\sigma+1)]}{[1]^{\sigma+1}}; \quad c^4 = \frac{[(m+1)(\sigma-1)]}{[1]^{\sigma+1}}. \tag{3.40}$$

Putting this information into the fourth powers of Equations 3.28 and 3.29 yields

$$d^4 = \frac{[m(\sigma-1)+(\sigma+1)]}{[1]^{\sigma+1}}; \quad b^4 = \frac{[(m-1)(\sigma-1)]}{[1]^{\sigma+1}}. \tag{3.41}$$

At this point we have completed a proof of the following theorem.

Theorem 3.10.3. *If there is a collineation of $GQ(\mathcal{C})$ that fixes the line $[A(\infty)]$ and the two points (∞) and $(0,0,0)$ and which is semilinear with associated field automorphism σ, there are two possibilities: $[A(0)]$ is mapped to $[A(\overline{0})]$ with $\overline{0} = \frac{[\sigma-1]}{[\sigma]}$ or $\overline{0} = \frac{[\sigma-1]}{[\sigma]} + [1]$. If the first case holds, then the collineation must be the following:*

$$\theta\left(\sigma, \begin{pmatrix} 1 & \left(\frac{[\sigma-1]}{[\sigma]}\right)^{\frac{1}{2}} \\ 0 & [1]^{\frac{1-\sigma}{2}} \end{pmatrix} \otimes \frac{1}{[1]^{\frac{\sigma+1}{4}}} \begin{pmatrix} [m(\sigma-1)-(\sigma+1)]^{\frac{1}{4}} & [(m-1)(\sigma-1)]^{\frac{1}{4}} \\ [(m+1)(\sigma-1)]^{\frac{1}{4}} & [m(\sigma-1)+(\sigma+1)]^{\frac{1}{4}} \end{pmatrix}\right).$$

According to the Fundamental Theorem, this will be a collineation of $GQ(\mathcal{C})$ if and only if $A_{\overline{t}} + A_{\overline{0}} = [1]^{1-\sigma} B^{-1} A_t^{(\sigma)} B^{-T}$, which in this case looks like

$$A_{[1]^{1-\sigma}t^\sigma + \frac{[\sigma-1]}{[\sigma]}} + A_{\frac{[\sigma-1]}{[\sigma]}} = [1]^{-\sigma} B^{-1} \begin{pmatrix} f(t)^\sigma & t^{\frac{\sigma}{2}} \\ 0 & g(t)^\sigma \end{pmatrix} B^{-T},$$

where $B^{-1} = \begin{pmatrix} \left[\frac{m(\sigma-1)+(\sigma+1)}{4}\right] & \left[\frac{(m-1)(\sigma-1)}{4}\right] \\ \left[\frac{(m+1)(\sigma-1)}{4}\right] & \left[\frac{m(\sigma-1)-(\sigma+1)}{4}\right] \end{pmatrix}$.

After multiplying out the above we find that we have proved the following under the additional assumption that an involution fixing $[A(\infty)]$ must be the one we call I_∞.

Theorem 3.10.4. *Suppose the line $[A(\infty)]$ is fixed by a unique nonidentity involution. Then there is a collineation of $GQ(\mathcal{C})$ that fixes the line $[A(\infty)]$ and the two points (∞) and $(0,0,0)$ and which is semi-linear with associated field automorphism σ if and only if the following two equations hold:*

$$f\left([1]^{1-\sigma}t^\sigma + \frac{[\sigma-1]}{[\sigma]}\right) + f\left(\frac{[\sigma-1]}{[\sigma]}\right) = [1]^{-\sigma}\left(\left[\frac{m(\sigma-1)+(\sigma+1)}{2}\right] f(t)^\sigma + \right.$$
$$\left.\left[\frac{m(\sigma-1)+(\sigma+1)}{4}\right]\left[\frac{(m-1)(\sigma-1)}{4}\right] t^{\frac{\sigma}{2}} + \left[\frac{(m-1)(\sigma-1)}{2}\right] g(t)^\sigma\right). \tag{3.42}$$

$$g\left([1]^{1-\sigma}t^{\sigma}+\frac{[\sigma-1]}{[\sigma]}\right)+g\left(\frac{[\sigma-1]}{[\sigma]}\right)=[1]^{-\sigma}\left(\left[\frac{(m+1)(\sigma-1)}{2}\right]f(t)^{\sigma}+\right.$$
$$\left.\left[\frac{(m+1)(\sigma-1)}{4}\right]\left[\frac{m(\sigma-1)-(\sigma+1)}{4}\right]t^{\frac{\sigma}{2}}+\left[\frac{m(\sigma-1)-(\sigma+1)}{2}\right]g(t)^{\sigma}\right). \tag{3.43}$$

These two equations hold if and only if the collineation of Theorem 3.10.3 actually is a collineation moving $[A(0)]$ to $[A(\frac{[\sigma-1]}{[\sigma]})]$. By composing with the involution $I_{\infty}=\theta_{-\frac{1}{2}}$ which maps $[A(t)]$ to $[A(t+\delta)]$, we see that both possible collineations of Theorem 3.10.3 exist or both fail to exist.

Note. It is clear that IF the two equations Eq. (3.42) and 3.43 both hold, then the two possible collineations of Theorem 3.10.3 both exist even without the hypothesis that there be a unique involution about the line $[A(\infty)]$. We just cannot say that these are the only possibilities.

As corollary to Theorem 3.10.3 we give the form of the collineation for $\sigma=2$.

Corollary 3.10.5. *If there is a collineation of $GQ(C)$ that fixes the line $[A(\infty)]$ and the two points (∞) and $(0,0,0)$ and which is semi-linear with associated field automorphism $\sigma=2$, there are two possibilities: $[A(0)]$ is mapped to $[A(\overline{0})]$ with $\overline{0}=\delta^{-1}$ or $\overline{0}=\delta^{-1}+\delta$. If the first case holds, then the collineation must be the following:*

$$\theta\left(2,\frac{1}{\delta^{\frac{1}{2}}}\begin{pmatrix}\delta^{\frac{1}{2}} & 1\\ 0 & 1\end{pmatrix}\otimes\frac{1}{[1]^{\frac{3}{4}}}\begin{pmatrix}\left[\frac{m-3}{4}\right] & \left[\frac{m-1}{4}\right]\\ \left[\frac{m+1}{4}\right] & \left[\frac{m+3}{4}\right]\end{pmatrix}\right).$$

Theorem 3.10.4 with $\sigma=2$ looks like:

Theorem 3.10.6. *Suppose the line $[A(\infty)]$ is fixed by a unique nonidentity involution. Then there is a collineation of $GQ(C)$ that fixes the line $[A(\infty)]$ and the two points (∞) and $(0,0,0)$ and which is semi-linear with associated field automorphism $\sigma=2$ if and only if the following two equations hold:*

$$f(\delta^{-1}t^2+\delta^{-1})+f(\delta^{-1})$$
$$=\frac{1}{\delta^2}\left(\left[\frac{m+3}{2}\right]f(t)^2+\left[\frac{m+3}{4}\right]\left[\frac{m-1}{4}\right]t+\left[\frac{m-1}{2}\right]g(t)^2\right). \tag{3.44}$$

$$g(\delta^{-1}t^2+\delta^{-1})+g(\delta^{-1})$$
$$=\frac{1}{\delta^2}\left(\left[\frac{m+1}{2}\right]f(t)^2+\left[\frac{m+1}{4}\right]\left[\frac{m-3}{4}\right]t+\left[\frac{m-3}{2}\right]g(t)^2\right). \tag{3.45}$$

We want to determine the order of the collineation given in Corollary 3.10.5. For this we don't need to carry along the entries of B. So write

$$\psi=\theta(2,\begin{pmatrix}1 & \delta^{-\frac{1}{2}}\\ 0 & \delta^{-\frac{1}{2}}\end{pmatrix}\otimes B)=\theta(2,A). \tag{3.46}$$

A little induction shows that

$$\psi^i = \theta(2^i, A^{(2^{i-1})} \cdot A^{(2^{i-2})} \cdots A) = \theta\left(2^i, \begin{pmatrix} 1 & \delta^{\frac{1}{2}}(\delta^{-1} + \delta^{-2} + \cdots (\delta^{-1})^{2^{i-1}}) \\ 0 & \delta^{\frac{1}{2}}(\delta^{-\frac{1}{2}})^{2^i} \end{pmatrix}\right). \tag{3.47}$$

Hence

$$\psi^e = \theta(id, \begin{pmatrix} 1 & \delta^{\frac{1}{2}} tr(\delta^{-1}) \\ 0 & 1 \end{pmatrix}), \tag{3.48}$$

which in any case must be I_∞. (Proof: By Theorem 2.5.4, any collineation of a non-classical $GQ(\mathcal{C})$ of the form $\theta(id, I \otimes B)$ must have $B = I$. Hence, if $\theta(id, A \otimes B_1)$ and $\theta(id, A \otimes B_2)$ are both collineations, then $B_1 = B_2$.) So ψ has order $2e$. Clearly $\varphi = \theta(id, P \otimes P)$ is in the group

$$\tilde{G} = \langle \tau_m = \theta(id, M),\ \psi = \theta\left(2, \begin{pmatrix} 1 & \delta^{-\frac{1}{2}} \\ 0 & \delta^{-\frac{1}{2}} \end{pmatrix}\right)\rangle.$$

It is also easy to check that

$$\psi^{-1} \circ \tau_m \circ \psi = {\tau_m}^2,$$

and

$$\langle \psi \rangle \cap \langle \tau_m \rangle = \{id\}.$$

Hence $\tilde{G} \cong C_{q+1} \rtimes C_{2e}$. Moreover, By L. E. Dickson [Di58] we know that $C_{q+1} \rtimes C_{2e}$ is a maximal subgroup of $P\Gamma L(2, q)$.

We collect the above observations about $\tilde{G}$ in the following:

Theorem 3.10.7. *Suppose that the conditions of Theorems* 3.8.1 *and* 3.10.4 *hold so that both* $\tau_m = \theta(id, M \otimes M^{-m})$ *and*

$$\psi = \theta\left(2, \frac{1}{\delta^{\frac{1}{2}}} \begin{pmatrix} \delta^{\frac{1}{2}} & 1 \\ 0 & 1 \end{pmatrix} \otimes \frac{1}{\delta^{\frac{3}{4}}} \begin{pmatrix} \left[\frac{m-3}{4}\right] & \left[\frac{m-1}{4}\right] \\ \left[\frac{m+1}{4}\right] & \left[\frac{m+3}{4}\right] \end{pmatrix}\right)$$

are collineations of $GQ(\mathcal{C})$. *Then*

$$\tilde{G} = \langle \tau_m, \psi \rangle \cong C_{q+1} \rtimes C_{2e}.$$

Also ψ *has order* $2e$ *and the flip* φ *is in* $\tilde{G}$. *Moreover,* $\tilde{G}$ *is a maximal subgroup of* $P\Gamma L(2, q)$.

Chapter 4

The Cyclic q-Clans

4.1 The Unified Construction of [COP03]

By a *cyclic q-clan* we mean one for which there is some m modulo $q+1$ for which the automorphism $\theta(id, M \otimes M^{-m})$ of $G^{\otimes}$ given explicitly by Eq. (3.21) is a collineation of $GQ(\mathcal{C})$. (See Theorem 3.8.1.) In [COP03] the authors gave a unified construction that included three previously known cyclic families plus a new one. We have modified their presentation to obtain what we call the *canonical* version. (See [Pa02a] for the connection between the original construction, which we do not need, and the one given here.) Moreover, we go on to show that the unified construction really does give cyclic GQ (see [CP03]).

First, recall the notation of Section 3.7. As before, ζ is a primitive element of $E = GF(q^2)$ and $\beta_k = \zeta^{k(q-1)}$, where $1 \le k \le q$. Then we have $[a] = \beta^a + \bar{\beta}^a$ and $[ka] = \beta_k^a + \bar{\beta}_k^a$. Write $T(x) = x + \bar{x}$ for $x \in E$ and note that $[k] = \beta_k + \bar{\beta}_k$. Put $\mathcal{Q} = \{\gamma \in E : \gamma^{q+1} = 1 \neq \gamma\} = \{\zeta^{k(q-1)} : k = 1, \ldots, q\}$. Define the following functions:

$$a_k(t) = \beta_k^{\frac{1}{2}} t + \bar{\beta}_k^{\frac{1}{2}}. \tag{4.1}$$

$$v_k(t) = t + (\delta_k t)^{\frac{1}{2}} + 1 = (t^{\frac{1}{2}} + \beta_k^{\frac{1}{2}})(t^{\frac{1}{2}} + \bar{\beta}_k^{\frac{1}{2}}). \tag{4.2}$$

$$h_1(t) = \frac{T\left((a_k(t))^m \bar{\beta}_k^{\frac{1}{2}}\right)}{v_k(t)^{m-1}}. \tag{4.3}$$

$$h_2(t) = \frac{T\left((a_k(t))^m \beta_k^{\frac{1}{2}}\right)}{v_k(t)^{m-1}}. \tag{4.4}$$

$$h_3(t) = \frac{T((t+\beta_k)^m)}{v_k(t)^{m-1}}. \tag{4.5}$$

$$M_t = \frac{t+\beta_k}{v_k(t)} = \frac{t^{\frac{1}{2}} + \beta_k^{\frac{1}{2}}}{t^{\frac{1}{2}} + \bar{\beta}_k^{\frac{1}{2}}}. \tag{4.6}$$

$$f_k(t) = \frac{\left[\frac{k(m+1)}{2}\right]}{[k]} + \frac{\left[\frac{k(m-1)}{2}\right]}{[k]}t + \frac{h_1(t)}{[k]} + \left(\frac{t}{[k]}\right)^{\frac{1}{2}}. \tag{4.7}$$

$$g_k(t) = \frac{\left[\frac{k(m-1)}{2}\right]}{[k]} + \frac{\left[\frac{k(m+1)}{2}\right]}{[k]}t + \frac{h_2(t)}{[k]} + \left(\frac{t}{[k]}\right)^{\frac{1}{2}}. \tag{4.8}$$

Let $\mathcal{C}_k = \left\{A_t = \begin{pmatrix} f_k(t) & t^{\frac{1}{2}} \\ 0 & g_k(t) \end{pmatrix} : t \in F\right\}$, where the functions $f_k(t)$ and $g_k(t)$ are the *canonical q-clan functions* of Eqs. (4.7) and (4.8).

Lemma 4.1.1. *Let ζ be a primitive element of E such that $\mathcal{Q} = \{\zeta^{k(q-1)}: k = 1, \ldots, q\}$. Then*

$$tr\left(\frac{1}{[km]}\right) = 1 \tag{4.9}$$

where $[km] = \beta_k^m + \bar{\beta}_k^m = \zeta^{k(q-1)m} + \bar{\zeta}^{k(q-1)m}$ (i.e., $\beta_k \in \mathcal{Q}$), and k is such that $q+1 \not| km$. In particular, if $(m, q+1) = 1$, then Eq. (4.9) holds for all $\beta_k \in \mathcal{Q}$.

Proof. First note that $tr\left(\frac{1}{[km]}\right) = 1$ if and only if the quadratic equation $x^2 + [km]x + 1$ is irreducible over $GF(q)$. This equation has roots β_k^m and $\bar{\beta}_k^m$, so $tr\left(\frac{1}{[km]}\right) = 1$ if and only if $\beta_k^m \notin GF(q)$. Now $\beta_k^m = \zeta^{km(q-1)} \in \mathcal{Q}$ is an element of $GF(q)$ if and only if $\zeta^{km(q-1)^2} = 1$. Hence if $\beta_k^m \in GF(q)$ then $(q+1)|km$, since $(q-1, q+1) = 1$. We have shown that if $\beta_k = \zeta^{k(q-1)}$ and $m \in \{1, ..., q^2-2\}$ satisfy $q+1 \not| km$, then $tr\left(\frac{1}{[km]}\right) = 1$, as required. Note that if $(m, q+1) = 1$ then $q+1 \not| km$. □

Theorem 4.1.2. *If m and k are nonzero residues modulo $q+1$, where $q+1$ does not divide km and $tr\left(\frac{T(\gamma^m)}{T(\gamma)}\right) = tr\left(\frac{[km]}{[k]}\right) = tr(1)$ for all $\gamma \in \mathcal{Q}$, then $\mathcal{C}_k$ is a q-clan.*

Proof. To prove the theorem we must show that for distinct $s, t \in F$ we have

$$tr\left(\frac{(f_k(s) + f_k(t))(g_k(s) + g_k(t))}{s+t}\right) = 1. \tag{4.10}$$

To aid in seeing the forest in spite of the trees we separate the following computations from the main proof. They are all completely routine.

$$a_k(t)\bar{a}_k(t) = v_k(t)^2. \tag{4.11}$$

$$v_k(t)^2 + v_k(s)^2 = s^2 + t^2 + [k](s+t). \tag{4.12}$$

$$\left[\frac{k(m+1)}{2}\right] h_1(t) + \left[\frac{k(m-1)}{2}\right] h_2(t) = [k]h_3(t). \tag{4.13}$$

$$h_1(t)h_2(t) = [k]v_k(t)^2 + \frac{h_1(t)^2 + h_2(t)^2}{[k]}. \tag{4.14}$$

$$\begin{aligned}\frac{h_1(t)h_2(t)+h_1(s)h_2(s)}{[2k](s+t)} &= \frac{v_k(t)^2+v_k(s)^2}{[k](s+t)}+\frac{h_1(t)^2+h_2(t)^2+h_1(s)^2+h_2(s)^2}{[k][2k](s+t)}\\ &= \frac{[k](s^2+t^2+[k](s+t))}{[2k](s+t)}+\frac{h_1(t)^2+h_2(t)^2+h_1(s)^2+h_2(s)^2}{[k][2k](s+t)}\\ &= \frac{s+t}{[k]}+1+\frac{h_1(t)^2+h_2(t)^2+h_1(s)^2+h_2(s)^2}{[k][2k](s+t)}. \end{aligned}\tag{4.15}$$

$$T(M_t)=\frac{[k]}{v_k(t)} \text{ and } \frac{h_3(t)}{[k]}=\frac{T(M_t^m)}{T(M_t)}.\tag{4.16}$$

$$T(M_s\bar{M}_t)=\frac{(s+t)[k]}{v_k(s)v_k(t)}.\tag{4.17}$$

$$h_1(t)h_2(s)+h_1(s)h_2(t)=[k]v_k(s)v_k(t)T(M_t^m\bar{M}_s^m).\tag{4.18}$$

Put

$$A=\frac{f_k(t)+f_k(s)}{(s+t)^{\frac{1}{2}}}=\frac{\left[\frac{k(m-1)}{2}\right]}{[k]}(t+s)^{\frac{1}{2}}+\frac{h_1(t)+h_1(s)}{[k](t+s)^{\frac{1}{2}}}+\frac{1}{\left[\frac{k}{2}\right]},$$

and put

$$B=\frac{g_k(t)+g_k(s)}{(s+t)^{\frac{1}{2}}}=\frac{\left[\frac{k(m+1)}{2}\right]}{[k]}(t+s)^{\frac{1}{2}}+\frac{h_2(t)+h_2(s)}{[k](t+s)^{\frac{1}{2}}}+\frac{1}{\left[\frac{k}{2}\right]}.$$

If we say $A=X+Y+Z$ and $B=X'+Y'+Z'$ in the obvious way, then we need to show that $1=tr(AB)$ where we write

$$AB=XX'+(XY'+YX')+(XZ'+X'Z)+YY'+(YZ'+Y'Z)+(ZZ').$$

We easily obtain the following terms:

$$XX'=\frac{(t+s)([km]+[k])}{[2k]}.\tag{4.19}$$

$$XY'+X'Y=\frac{[k](h_3(t)+h_3(s))}{[2k]}=\frac{T(M_t^m)}{T(M_t)}+\frac{T(M_s^m)}{T(M_s)}.\tag{4.20}$$

$$XZ'+X'Z=\frac{(t+s)^{\frac{1}{2}}\left[\frac{km}{2}\right]\left[\frac{k}{2}\right]}{[k]\left[\frac{k}{2}\right]}.\tag{4.21}$$

$$\begin{aligned}YY' &= \frac{h_1(t)h_2(t)+h_1(s)h_2(s)}{[2k](s+t)}+\frac{h_1(t)h_2(s)+h_1(s)h_2(t)}{[2k](s+t)}\\ &= \left\{\frac{s+t}{[k]}+1+\frac{h_1(t)^2+h_2(t)^2+h_1(s)^2+h_2(s)^2}{[k][2k](s+t)}\right\}+\frac{T(M_t^m\bar{M}_s^m)}{T(M_t\bar{M}_s)}.\end{aligned}\tag{4.22}$$

$$YZ' + Y'Z = \frac{h_1(t) + h_2(t) + h_1(s) + h_2(s)}{(s+t)^{\frac{1}{2}[k]\left[\frac{k}{2}\right]}}. \tag{4.23}$$

$$ZZ' = \frac{1}{[k]}. \tag{4.24}$$

We can lump into a general term C whose absolute trace is zero any pair of terms of the form $t + t^2$. Clearly XX' breaks into a sum of two terms, the first of which is the square of $XZ' + X'Z$, the second of which cancels with one of the terms in YY'. Another summand of YY' is the square of $YZ' + Y'Z$. When adding up the remaining terms and noticing that M_t and $M_t\bar{M}_s$ belong to $\mathcal{Q}$, we find that

$$AB = C + \frac{T(M_t^m)}{T(M_t)} + \frac{T(M_s^m)}{T(M_s)} + 1 + \frac{T(M_t^m\bar{M}_s^m)}{T(M_t\bar{M}_s)} + \frac{1}{[k]}, \tag{4.25}$$

so that $tr(AB) = 0 + tr(1) + tr(1) + tr(1) + tr(1) + 1 = 1$, as needed. □

Observation 4.1.3. Note that if $m \equiv 5 \pmod{q+1}$ and $e \equiv 2 \pmod 4$, there are four values of k (i.e., $k = j(q+1)/5$, $1 \le j \le 4$), for which $q+1|5k$ (and so the hypothesis of Theorem 4.1.2 is not satisfied) but it still holds that $tr(1/[k]) = 1$ in Eq. (4.25). In the proof of Theorem 4.1.2, the only place in which we use the fact that $q+1$ does not divide km is in the last term of Eq. (4.25) (more precisely, we need to apply Lemma 4.1.1 to show that $tr(1/[k]) = 1$). However, in the case $m \equiv 5 \pmod{q+1}$ and $e \equiv 2 \pmod 4$, we have $tr(1/[k]) = 1$, hence Eq. (4.25) is satisfied and $\mathcal{C}_k$ is a q-clan.

Observation 4.1.4. Notice that if either k or m is relatively prime to $q+1$, the condition that $q+1$ does not divide km is satisfied. In particular, we see that $k = 1$ will always satisfy the condition.

Lemma 4.1.5. *Interchanging m and $-m$ amounts to interchanging $f_k(t)$ and $g_k(t)$ in the q-clan matrices.*

Proof. Suppose that $\mathcal{C}_k = \left\{A_t = \begin{pmatrix} f_k(t) & t^{\frac{1}{2}} \\ 0 & g_k(t) \end{pmatrix} : t \in F\right\}$ is a q-clan for which $\theta = \theta(id, M \otimes M^{-m})$ is a collineation of $GQ(C)$ (acting as a cycle of length $q+1$ on the lines through (∞)). In particular θ is an automorphism of $G^{\otimes}$. Also, $\varphi = \theta(id, I \otimes P)$ is an involutory automorphism of $G^{\otimes}$ mapping $(\gamma \otimes \alpha, c)$ to $(\gamma \otimes \alpha P, c)$, i.e., it leaves each $\mathcal{L}_\gamma$ invariant.

Also,

$$\varphi : (\gamma_{t^{\frac{1}{2}}} \otimes \alpha, \alpha A_t \alpha^T) \mapsto (\gamma_{t^{\frac{1}{2}}} \otimes \alpha P, (\alpha P)(PA_tP)(\alpha P)^T), \; t \in \tilde{F}.$$

So φ is an automorphism of $G^{\otimes}$ for which $\varphi : GQ(\mathcal{C}_k) \to GQ(\mathcal{C}'_k)$, where $\mathcal{C}'_k = \left\{PA_tP \equiv \begin{pmatrix} g_k(t) & t^{\frac{1}{2}} \\ 0 & f_k(t) \end{pmatrix} : t \in F\right\}$, and $A'_\infty = A_\infty = \left(\begin{smallmatrix} 0 & 0 \\ 0 & 0 \end{smallmatrix}\right)$. Then, since $PM^{-m}P = M^m$, $\varphi^{-1} \circ \theta \circ \varphi$ is an automorphism of $G^{\otimes}$ for which

$$\begin{aligned}\varphi^{-1} \circ \theta \circ \varphi &= \theta(id, I \otimes P) \circ \theta(id, M \otimes M^{-m}) \circ \theta(id, I \otimes P) \\ &= \theta(id, M \otimes PM^{-m}P) = \theta(id, M \otimes M^m)\end{aligned}$$

is a collineation of $GQ(\mathcal{C}'_k)$ which acts as a cycle of length $q+1$ on the lines through the point $(\infty)'$. It follows that replacing m with $-m$ is equivalent to interchanging $f_k(t)$ with $g_k(t)$, which gives an isomorphic GQ. □

4.2 The Known Cyclic q-Clans

In [COP03] there is the following remarkable theorem.

Theorem 4.2.1. *Let k be a nonzero residue mod $q+1$. Four infinite families of q-clans arise:*

(1) *If $m \equiv \pm 1 \pmod{q+1}$, then $\mathcal{C}$ is the classical q-clan for all $q = 2^e$ and all k. $GQ(\mathcal{C})$ is isomorphic to $H(3, q^2)$.*

(2) *If $m \equiv \pm\frac{q}{2} \pmod{q+1}$ and e is odd, then $\mathcal{C}_k$ is the FTWKB q-clan for all k, i.e., in standard form*[1] $A_t \equiv \begin{pmatrix} t^{\frac{1}{4}} & t^{\frac{2}{3}} \\ 0 & t^{\frac{1}{4}} \end{pmatrix}$. *(see Eq. (5.6)).*

(3.a) *If $m \equiv \pm 5 \pmod{q+1}$ and $e \not\equiv 2 \pmod 4$, then $\mathcal{C}_k$ is the Subiaco q-clan for all k.*

(3.b) *If $m \equiv \pm 5 \pmod{q+1}$, $e \equiv 2 \pmod 4$ and $q+1 \not| 5k$, then $\mathcal{C}_k$ is the Subiaco q-clan.*

(3.c) *If $m \equiv \pm 5 \pmod{q+1}$ and $e \equiv 2 \pmod 4$, there are four values of k (i.e., $k = j(q+1)/5$, $1 \le j \le 4$), for which $q+1 | 5k$ but $\mathcal{C}_k$ is still a Subiaco q-clan.*

(4) *If $m \equiv \pm\left(\frac{q-1}{3}\right) \pmod{q+1}$ with e even, then for all k, $\mathcal{C}$ is a new q-clan called the Adelaide q-clan.*

Proof. For each nonzero residue k modulo $q+1$ we must show that for the appropriate choice of e and m we have

$$tr\left(\frac{[km]}{[k]}\right) = tr(1). \tag{4.26}$$

First, for all e and $m = 1$ it is clear that Eq. (4.26) holds. By Theorem 4.1.2, $\mathcal{C}$ is a q-clan. The resulting GQ is classical and isomorphic to $H(3, q^2)$. For a lengthy discussion of this example see Section 5.1. This takes care of (1).

[1]by standard form we mean the form of equations as in [COP03].

Second, for all odd e put $m = -\frac{q}{2} \equiv \frac{1}{2} \pmod{q+1}$. Then using Eq. (3.17) we have

$$tr\left(\frac{[km]}{[k]}\right) = tr\left(\frac{1}{\left[\frac{k}{2}\right]}\right) = 1 = tr(1).$$

Applying Theorem 4.1.2, case (2) follows and gives the FTWKB example. See Section 5.2 for a discussion of these GQ.

For all e put $m = 5$. Using part (vi) of Lemma 3.7.1 we have

$$tr\left(\frac{[5k]}{[k]}\right) = tr([k]^4 + [k]^2 + 1) = tr(1). \tag{4.27}$$

By Theorem 4.1.2, case $(3.a)$ immediately follows and (using Lemma 4.1.1) also case $(3.b)$ easily follows. Now suppose $m \equiv \pm 5 \pmod{q+1}$ and $e \equiv 2 \pmod 4$. Since $e \equiv 0 \pmod 2$, $GF(4) \leq GF(q)$, so there is an element $\omega \in GF(q)$ with $\omega^2 + \omega + 1 = 0$. It follows that $tr(\omega) = 1$. For $k = j(q+1)/5$, $1 \leq j \leq 4$, we have $[k] \in \{\omega, \omega^2\}$. In this case $q+1|km$, but $\mathcal{C}_k$ is still a Subiaco q-clan because Eq. (4.27) holds and it is still true that $tr(1/[k]) = tr(\omega^2) = 1$. (See Observation 4.1.3.) This completes case $(3.c)$.

Finally, for even e, so $tr(1) = 0$, put $m = \frac{q-1}{3} \equiv -\frac{2}{3} \pmod{q+1}$. Again using part (vi) of Lemma 3.7.1 we have

$$\begin{aligned} tr\left(\frac{\left[\frac{-2k}{3}\right]}{[k]}\right) &= tr\left(\frac{\left[\frac{2k}{3}\right]}{\left[\frac{k}{3}\right]^3 + \left[\frac{k}{3}\right]}\right) \\ &= tr\left(\frac{\left[\frac{k}{3}\right] + 1 + 1}{\left[\frac{k}{3}\right]^2 + 1}\right) = tr\left(\frac{1}{\left[\frac{k}{3}\right] + 1} + \left(\frac{1}{\left[\frac{k}{3}\right] + 1}\right)^2\right) = 0 = tr(1). \end{aligned}$$

This shows that a GQ is obtained by Theorem 4.1.2. For a fuller treatment of these GQ, named the Adelaide GQ in [COP03], see Section 5.4. □

4.3 q-Clan Functions Via the Square Bracket

For a fixed k, let f_k and g_k denote the corresponding q-clan functions. We want to express $f_k(t)$ and $g_k(t)$ in terms of the bracket function, starting with Eq. (4.7).

Then with $t = \frac{[j+k]}{[j]}$ Eq. (4.7) says

$$f_k(t) = f_k\left(\frac{[j+k]}{[j]}\right) = \frac{\left[k\left(\frac{m+1}{2}\right)\right][j]}{[k][j]} + \frac{\left[k\left(\frac{m-1}{2}\right)\right][j+k]}{[k][j]}$$
$$+ \frac{(\beta_k^{\frac{1}{2}}t + \overline{\beta}_k^{\frac{1}{2}})^m \overline{\beta}_k^{\frac{1}{2}} + (\overline{\beta}_k^{\frac{1}{2}}t + \beta_k^{\frac{1}{2}})^m \beta_k^{\frac{1}{2}}}{[k](t^{\frac{1}{2}} + \beta_k^{\frac{1}{2}})^{m-1}(t^{\frac{1}{2}} + \overline{\beta}_k^{\frac{1}{2}})^{m-1}} + \left(\frac{[j+k]}{[j][k]}\right)^{\frac{1}{2}}$$
$$= \frac{\left[k\left(\frac{m+1}{2}\right) + j\right] + \left[k\left(\frac{m+1}{2}\right) - j\right] + \left[k\left(\frac{m+1}{2}\right) + j\right] + \left[k\left(\frac{m-1}{2}\right) - j - k\right]}{[k][j]}$$
$$+ T + \left(\frac{[j+k]}{[j][k]}\right)^{\frac{1}{2}}$$
$$= \frac{\left[k\left(\frac{m+1}{2}\right) - j\right] + \left[k\left(\frac{m-1}{2}\right) - j - k\right]}{[k][j]} + T + \left(\frac{[j+k]}{[j][k]}\right)^{\frac{1}{2}}.$$

Here we can rewrite T as

$$T = \frac{\beta_k^{\frac{m-1}{2}}(t + \overline{\beta}_k)^{\frac{m+1}{2}}}{[k](t+\beta_k)^{\frac{m-1}{2}}} + \frac{\overline{\beta}_k^{\frac{m-1}{2}}(t + \beta_k)^{\frac{m+1}{2}}}{[k](t+\overline{\beta}_k)^{\frac{m-1}{2}}}$$
$$= \frac{\beta_k^{\frac{m-1}{2}}(t + \overline{\beta}_k)}{[k]}\left(\frac{t+\overline{\beta}_k}{t+\beta_k}\right)^{\frac{m-1}{2}} + \frac{\overline{\beta}_k^{\frac{m-1}{2}}(t + \beta_k)}{[k]}\left(\frac{t+\beta_k}{t+\overline{\beta}_k}\right)^{\frac{m-1}{2}}.$$

It is routine to show that for $t = \frac{[j+k]}{[j]}$, $\left(\frac{t+\overline{\beta}_k}{t+\beta_k}\right) = \beta^{2j}$ and $\left(\frac{t+\beta_k}{t+\overline{\beta}_k}\right) = \overline{\beta}^{2j}$. Then after a few steps

$$T = \frac{\left[jm + k\left(\frac{m-1}{2}\right)\right]}{[j]}.$$

Hence

Theorem 4.3.1.

$$f_k\left(\frac{[j+k]}{[j]}\right) = \frac{\left[k\left(\frac{m-1}{2}\right) - j\right][k]}{[j][k]} + \frac{\left[jm + k\left(\frac{m-1}{2}\right])\right]}{[j]} + \left(\frac{[j+k]}{[j][k]}\right)^{\frac{1}{2}}$$
$$= \frac{\left[(k+j)\left(\frac{m-1}{2}\right)\right]\left[j\left(\frac{m+1}{2}\right)\right]}{[j]} + \left(\frac{[j+k]}{[j][k]}\right)^{\frac{1}{2}}.$$

In a very similar sequence of steps we find that

$$g_k(t) = g_k\left(\frac{[j+k]}{[j]}\right) = \frac{\left[(j+k)\left(\frac{m+1}{2}\right)\right]\left[j\left(\frac{m-1}{2}\right)\right]}{[j]} + \left(\frac{[j+k]}{[j][k]}\right)^{\frac{1}{2}}.$$

Using these forms of f_k and g_k we can study the collineations and isomorphisms of the cyclic GQ. Also, note that when f_k and g_k are given in this form, Lemma 4.1.5 is a triviality! Moreover, replacing m by any integer congruent to m modulo $q+1$ leaves the q-clan functions unchanged. This is not at all obvious from the form of the functions given in Eqs. (4.7) and (4.8).

4.4 The Flip is a Collineation

Theorem 4.4.1. *The q-clan functions f_k and g_k are reverses of each other, i.e., $tg_k\left(\frac{1}{t}\right) = f_k(t)$. Hence by Theorem 3.2.1 we have*

$$\theta(id, P \otimes P)\colon (\alpha, \beta, c) \mapsto ((\alpha, \beta)(P \otimes P), c + \alpha \circ \beta)$$

is the involution $I_1^{(k)} : [A(t)]_k \leftrightarrow [A(t^{-1})]_k$ for $t \in \tilde{F}$.

Here we adopt a natural indexing of objects to indicate the choice of k. Since $I_1^{(k)}$ is the involution of $GQ(\mathcal{C}_k)$ fixing the line $[A(1)]_k$, using (k) as a superscript to identify the GQ on which $I_1^{(k)}$ acts should not lead to confusion.

Proof.

$$tg_k\left(\frac{1}{t}\right) = \left(\frac{[j+k]}{[j]}\right) g_k\left(\frac{[j]}{[j+k]}\right) = \frac{[j+k]}{[j]} g_k\left(\frac{[-j-k+k]}{[-j-k]}\right)$$
$$= \frac{[j+k]}{[j]}\left\{\frac{[(-j)\left(\frac{m+1}{2}\right)][(-j-k)\left(\frac{m-1}{2}\right)]}{[-j-k]} + \left(\frac{[-j]}{[-j-k][k]}\right)^{\frac{1}{2}}\right\}$$
$$= \frac{[j\left(\frac{m+1}{2}\right)][(j+k)\left(\frac{m-1}{2}\right)]}{[j]} + \left(\frac{[j+k]}{[j][k]}\right)^{\frac{1}{2}} = f_k\left(\frac{[j+k]}{[j]}\right).$$

□

Theorem 4.4.2. *Let $A = \begin{pmatrix} 1 & [\frac{k}{2}] \\ 0 & 1 \end{pmatrix}$ and $B = \frac{1}{[\frac{k}{2}]}\begin{pmatrix} [\frac{km}{2}] & [\frac{k(m-1)}{2}] \\ [k(\frac{m+1}{2})] & [\frac{km}{2}] \end{pmatrix}$. Then $\theta(id, A \otimes B) : ((\alpha, \beta), c) \mapsto ((\alpha, \beta)(A \otimes B), c + \alpha B A_k B^T \alpha^T)$ is the involution $I_\infty^{(k)}$ mapping $[A(t)]_k$ to $[A(t + [k])]_k$ for $t \in F$ and fixing $[A(\infty)]$.*

Proof. The details of showing that $A_{\bar{t}} + A_{\bar{0}} \equiv B^{-1} A_t B^{-T}$ (where all the matrices use q-clan functions for a fixed but arbitrary k) are routine. We leave them as an exercise for the reader. This is partly justified because the results in the next two sections make this result unnecessary. □

It is also a routine exercise to calculate

$$I_1^{(k)} \circ I_\infty^{(k)}$$
$$= \theta\left(id, \begin{pmatrix} 0 & 1 \\ 1 & [\frac{k}{2}] \end{pmatrix} \otimes \frac{1}{[\frac{k}{2}]}\begin{pmatrix} [k(\frac{m+1}{2})] & [\frac{km}{2}] \\ [\frac{km}{2}] & [k(\frac{m-1}{2})] \end{pmatrix}\right) = \theta(id, M \otimes M^{-m}).$$

It follows that if we had given a proof of Theorem 4.4.2 we would not have to give the proof of Theorem 4.6.1 that appears in Section 4.6. However, the two proofs are not that different, and we felt obliged to give at least one of them.

4.5 The Main Isomorphism Theorem

The next result shows that we may put $k = 1$ from now on. Note: This proof is ever so much simpler than the proof in the SN that any two Subiaco GQ of the same order are isomorphic.

Let $\mathcal{C}$ denote the q-clan obtained by putting $k = 1$ and let $\mathcal{C}_k$ denote the corresponding q-clan for arbitrary k.

Theorem 4.5.1. *Let* $\mu = \frac{\left[\frac{k}{2}\right]}{\left[\frac{1}{2}\right]} = det(A)$ *where we put*

$$A = \begin{pmatrix} 1 & \frac{\left[\frac{k-1}{2}\right]}{\left[\frac{1}{2}\right]} \\ 0 & \frac{\left[\frac{k}{2}\right]}{\left[\frac{1}{2}\right]} \end{pmatrix}, \quad \textit{and } B = \frac{1}{\left[\frac{1}{4}\right]\left[\frac{k}{4}\right]} \begin{pmatrix} \left[\frac{(k-1)(m-1)+2k}{4}\right] & \left[\frac{(k-1)(m-1)}{4}\right] \\ \left[\frac{(k-1)(m+1)}{4}\right] & \left[\frac{(k-1)(m-1)-2}{4}\right] \end{pmatrix}.$$

Then $\theta(id, A \otimes B)$ *is an isomorphism from* $GQ(\mathcal{C})$ *to* $GQ(\mathcal{C}_k)$ *mapping* $[A(\infty)]$ *to* $[A(\infty)]_k$ *and in general mapping* $[A(t)]$ *to* $[A(\bar{t})]_k$ *where* $\bar{t} = \frac{[k]}{[1]}t + \frac{[k-1]}{[1]}$. *More important for computations, if* $t = \frac{[j+1]}{[j]}$, *then* $\bar{t} = \frac{[j+k]}{[j]}$.

Proof. By the F.T. we must show that if $A'_{\bar{t}} = \begin{pmatrix} f_k(t) & t^{\frac{1}{2}} \\ 0 & g_k(t) \end{pmatrix}$ and A_t is the corresponding q-clan matrix when $k = 1$, then

$$A'_{\bar{t}} + A'_{\bar{0}} \equiv \mu B^{-1} A_t B^{-T}. \tag{4.28}$$

First check that $det(B) = 1$, so that

$$B^{-1} = \frac{1}{\left[\frac{1}{4}\right]\left[\frac{k}{4}\right]} \begin{pmatrix} \left[\frac{(k-1)(m-1)-2}{4}\right] & \left[\frac{(k-1)(m-1)}{4}\right] \\ \left[\frac{(k-1)(m+1)}{4}\right] & \left[\frac{(k-1)(m-1)+2k}{4}\right] \end{pmatrix}.$$

Using Observation 1.1.3 we easily compute that $\mu B^{-1} A_t B^{-T} \equiv \frac{1}{[1]}\begin{pmatrix} x & y \\ z & w \end{pmatrix}$, where

$$x = \left[\frac{(k-1)(m-1)-2}{2}\right] f(t) + \left[\frac{(k-1)m-1)-2}{4}\right]\left[\frac{(k-1)(m-1)}{4}\right] t^{\frac{1}{2}} + \left[\frac{(k-1)(m-1)}{2}\right] g(t),$$

$$y = \left[\frac{k}{2}\right]\left[\frac{1}{2}\right] t^{\frac{1}{2}},$$

$$z = 0,$$

$$w = \left[\frac{(k-1)(m+1)}{2}\right] f(t) + \left[\frac{(k-1)(m+1)}{4}\right]\left[\frac{(k-1)(m-1)+2k}{4}\right] t^{\frac{1}{2}} + \left[\frac{(k-1)(m-1)+2k}{2}\right] g(t).$$

It is quite easy to check that the entry in the (1,2) position on left-hand side of Eq. (4.28) equals the entry in the (1,2) position on the right-hand side. It is not so easy to check the entries in the (1,1) positions and the (2,2) positions, but the details are quite similar for the two positions. We give fairly complete details for the (1,1) positions and leave the details for the other case as an exercise for the reader.

First consider the left-hand side. So

$$\begin{aligned}
f_k(\bar{t}) + f_k(\bar{0}) &= f_k\left(\frac{[j+k]}{[j]}\right) + f_k\left(\frac{[-1+k]}{[-1]}\right) \\
&= \frac{[(k+j)\left(\frac{m-1}{2}\right)]\,[j\left(\frac{m+1}{2}\right)]}{[j]} + \left(\frac{[j+k]}{[j][k]}\right)^{\frac{1}{2}} \\
&\quad + \frac{[(k-1)\left(\frac{m-1}{2}\right)]\,[\frac{m+1}{2}]}{[1]} + \left(\frac{[k-1]}{[1][k]}\right)^{\frac{1}{2}} \\
&= \frac{[(k+j)\left(\frac{m-1}{2}\right)]\,[j\left(\frac{m+1}{2}\right)]\,[1] + [(k-1)\left(\frac{m-1}{2}\right)]\,[\frac{m+1}{2}]\,[j]}{[j][1]} \\
&\quad + \left(\frac{1}{[k]}\left(\frac{[j+k][1]+[k-1][j]}{[j][1]}\right)\right)^{\frac{1}{2}} \\
&= \frac{[k\left(\frac{m-1}{2}\right)+jm+1] + [k\left(\frac{m-1}{2}\right)+jm-1]}{[j][1]} \\
&\quad + \frac{[k\left(\frac{m-1}{2}\right)]\,[j+1] + [k\left(\frac{m-1}{2}\right)-m+j]}{[j][1]} \\
&\quad + \frac{[k\left(\frac{m-1}{2}\right)-m-j]}{[j][1]} + \left(\frac{[j+1]}{[j][1]}\right)^{\frac{1}{2}}.
\end{aligned}$$

Now for the entry x in the (1,1) position of the right-hand side we obtain

$$\begin{aligned}
&\frac{1}{[1]}\left[\frac{(k-1)(m-1)-2}{2}\right]\left\{\frac{[(j+1)\left(\frac{m-1}{2}\right)][j\left(\frac{m+1}{2}\right)]}{[j]}+\left(\frac{[j+1]}{[j][1]}\right)^{\frac{1}{2}}\right\}\\
&\qquad+\frac{\left[\frac{(k-1)(m-1)-2}{4}\right]\left[\frac{(k-1)(m-1)}{4}\right]}{[1]}\left(\frac{[j+1]}{[j]}\right)^{\frac{1}{2}}\\
&\qquad+\frac{\left[\frac{(k-1)(m-1)}{2}\right]}{[1]}\left\{\frac{[(j+1)\left(\frac{m+1}{2}\right)][j\left(\frac{m-1}{2}\right)]}{[j]}+\left(\frac{[j+1]}{[j][1]}\right)^{\frac{1}{2}}\right\}\\
&=\frac{\left[\frac{(k-1)(m-1)-2}{2}\right][(j+1)\left(\frac{m-1}{2}\right)][j\left(\frac{m+1}{2}\right)]}{[1][j]}+\\
&\qquad+\frac{\left[\frac{(k-1)(m-1)}{2}\right][(j+1)\left(\frac{m+1}{2}\right)][j\left(\frac{m-1}{2}\right)]}{[1][j]}\\
&\qquad+\left(\frac{[j+1]}{[j][1]}\right)^{\frac{1}{2}}\left\{\frac{\left[\frac{(k-1)(m-1)-2}{2}\right]+\left[\frac{(k-1)(m-1)-2}{4}\right]}{[1]}\right\}\\
&\qquad\qquad\times\left\{\frac{\left[\frac{(k-1)(m-1)}{4}\right]\left[\frac{1}{2}\right]+\left[\frac{(k-1)(m-1)}{2}\right]}{[1]}\right\}.
\end{aligned}$$

First consider the coefficient on the term $\left(\frac{[j+1]}{[j][1]}\right)^{\frac{1}{2}}$. If we write $a=\frac{(k-1)(m-1)}{2}$, then this coefficient has the form $\frac{[a-1]+[\frac{a-1}{2}][\frac{a}{2}][\frac{-1}{2}]+[a]}{[1]}=1$, which equals the corresponding coefficient on the left-hand side. For the remaining terms, continue to use $a=\frac{(k-1)(m-1)}{2}$, so $\frac{k(m-1)}{2}=a+\frac{m}{2}-\frac{1}{2}$. Expand out the products on the right-hand side and multiply through by the denominator $[1][j]$. Two terms cancel, leaving the following six terms:

$$[a+\frac{m}{2}+jm+\frac{1}{2}]+[a+\frac{m}{2}+jm-\frac{3}{2}]+[a+\frac{m}{2}-j-\frac{3}{2}]$$

$$+[a+\frac{m}{2}+j+\frac{1}{2}]+[a-\frac{m}{2}-j-\frac{1}{2}]+[a-\frac{m}{2}+j-\frac{1}{2}].$$

When the terms on the left-hand side above are expanded in the same fashion, exactly the same terms appear, completing this part of the proof. □

4.6 The Unified Construction Gives Cyclic q-Clans

From now on we assume that $k = 1$, i.e., $\beta = \beta_k$ has multiplicative order $q+1$. The theorem of this section shows that every GQ arising from the unified construction is cyclic in the sense that it admits the collineation $\theta(id, M \otimes M^{-m})$ which permutes the lines through (∞) in a cycle of length $q+1$.

Theorem 4.6.1 ([CP03]). *The q-clan functions of Eqs.* (4.7) *and* (4.8) *(with $k = 1$) satisfy the two conditions of Theorem* 3.8.1.

Proof. To show that the two conditions of Theorem 3.8.1 are both satisfied is routine but a bit tedious. The two verifications are quite similar, so we give only the first one.

For the first condition we need to show that

$$[1]t\left\{f(t^{-1}+\delta)+f(\delta)\right\} = [m-1]f(t) + \left[\frac{m-1}{2}\right]\left[\frac{m}{2}\right]t^{\frac{1}{2}} + [m]g(t). \qquad (4.29)$$

Start with the left-hand side and do some special pieces of the computation first: $a(\delta)^m\bar{\beta}^{\frac{1}{2}} = \beta^{\frac{3m-1}{2}}$; $v(\delta) = 1$; $tr(a(\delta)^m\bar{\beta}^{\frac{1}{2}}) = \left[\frac{3m-1}{2}\right]$. So $[1]f(\delta) = \left[\frac{m+1}{2}\right] + \left[\frac{m-1}{2}\right][1] + \left[\frac{3m-1}{2}\right] + [1]$.

Next, compute $v(t^{-1}+\delta) = v(t)t^{-1}$; $a(t^{-1}+\delta) = \beta a(t)t^{-1}$, so

$$\frac{T(a(t^{-1}+\delta)^m\bar{\beta}^{\frac{1}{2}})}{v(t^{-1}+\delta)^{m-1}} = \frac{T\left(a(t)^m\beta^{m-\frac{1}{2}}\right)}{tv(t)^{m-1}}.$$

Then with several terms cancelling we find:

$$[1]t\left\{f(t^{-1}+\delta)+f(\delta)\right\} = \left[\frac{m-1}{2}\right] + \left[\frac{3m-1}{2}\right]t + \frac{T(a(t)^m\beta^{m-\frac{1}{2}})}{v(t)^{m-1}} + (t\delta)^{\frac{1}{2}}. \qquad (4.30)$$

Now we begin with the right-hand side of Eq. (4.29) which is

$$\frac{[m-1]}{[1]}\left\{\left[\frac{m+1}{2}\right] + \left[\frac{m-1}{2}\right]t + \frac{T(a(t)^m\bar{\beta}^{\frac{1}{2}})}{v(t)^{m-1}} + (t\delta)^{\frac{1}{2}}\right\} + \left[\frac{m-1}{2}\right]\left[\frac{m}{2}\right]t^{\frac{1}{2}}$$
$$+\frac{[m]}{[1]}\left\{\left[\frac{m-1}{2}\right] + \left[\frac{m+1}{2}\right]t + \frac{T(a(t)^m\beta^{\frac{1}{2}})}{v(t)^{m-1}} + (t\delta)^{\frac{1}{2}}\right\}.$$

Consider the summands one at a time. First, the constant term equals

$$\frac{\left[\frac{2m-2}{2}\right]\left[\frac{m+1}{2}\right] + \left[\frac{2m}{2}\right]\left[\frac{m-1}{2}\right]}{[1]} = \left[\frac{m-1}{2}\right].$$

The coefficient on t equals

$$\frac{\left[\frac{2m-2}{2}\right]\left[\frac{m-1}{2}\right]+\left[\frac{2m}{2}\right]\left[\frac{m+1}{2}\right]}{[1]}=\left[\frac{3m-1}{2}\right].$$

The coefficient on $t^{\frac{1}{2}}$ is

$$\frac{[m-1]+\left[\frac{m-1}{2}\right]\left[\frac{m}{2}\right]\left[\frac{1}{2}\right]+[m]}{\left[\frac{1}{2}\right]}=\left[\frac{1}{2}\right].$$

Finally, we consider the term

$$\begin{aligned}\frac{[m-1]T(a(t)^m\bar{\beta}^{\frac{1}{2}})+[m]T(a(t)^m\beta^{\frac{1}{2}})}{[1]v(t)^{m-1}} &= \frac{T\left\{a(t)^m(\beta^{m-1}+\bar{\beta}^{m-1})\bar{\beta}^{\frac{1}{2}}+a(t)^m(\beta^m(\beta^m+\bar{\beta}^m)\beta^{\frac{1}{2}}\right\}}{[1]v(t)^{m-1}}\\ &= \frac{T(a(t)^m\beta^{m-\frac{1}{2}})}{v(t)^{m-1}},\end{aligned}$$

with another step or two.

When we compare the left- and right-hand sides, we see that we have equality. □

At this point we know that the involutions studied in Section 3.9 are all collineations of $GQ(\mathcal{C})$.

4.7 Some Semi-linear Collineations

In this section we show that the cyclic q-clans of the unified construction admit the semi-linear collineations described in Corollary 3.10.5, i.e., the q-clan functions satisfy the conditions of Eqs. (3.44) and (3.45) in Theorem 3.10.4.

Theorem 4.7.1. *There is a collineation of $GQ(\mathcal{C})$ that fixes the line $[A(\infty)]$ and the two points (∞) and $(0,0,0)$ and which is semilinear with associated field automorphism $\sigma = 2$. Specifically, the q-clan functions satisfy the conditions of Eqs.* (3.44) *and* (3.45). *In this case the collineation must be the following:*

$$\theta\left(2,\frac{1}{\delta^{\frac{1}{2}}}\begin{pmatrix}\delta^{\frac{1}{2}} & 1\\ 0 & 1\end{pmatrix}\otimes\frac{1}{\delta^{\frac{1}{2}}}\begin{pmatrix}\frac{\left[\frac{m-3}{4}\right]}{\left[\frac{1}{4}\right]} & \frac{\left[\frac{m-1}{4}\right]}{\left[\frac{1}{4}\right]}\\ \frac{\left[\frac{m+1}{4}\right]}{\left[\frac{1}{4}\right]} & \frac{\left[\frac{m+3}{4}\right]}{\left[\frac{1}{4}\right]}\end{pmatrix}\right).$$

Proof. We give the details only for the Eq. (3.45).

$$
\begin{aligned}
& g\left(\frac{t^2+1}{\delta}\right)+g\left(\frac{1}{\delta}\right) \\
&= \frac{\left[\frac{m+1}{2}\right]}{[1]}\left(\frac{t^2+1+1}{[1]}\right)+\frac{T((a\left(\frac{t^2+1}{\delta}\right))^m\beta^{\frac12})}{[1]\left(\frac{t^2+1}{\delta}+(t+1)+1\right)^{m-1}}+\frac{T((a\left(\frac{1}{\delta}\right))^m\beta^{\frac12})}{[1]\left(\frac{1}{\delta}+1+1\right)^{m-1}}+\frac{t}{\delta} \\
&= \frac{\left[\frac{m+1}{2}\right]t^2}{[2]}+\frac{t}{\delta}+\frac{\left(\beta^{\frac12}\frac{1}{\delta}+\overline{\beta}^{\frac12}\right)^m\beta^{\frac12}+\left(\overline{\beta}^{\frac12}\frac{1}{\delta}+\beta^{\frac12}\right)^m\overline{\beta}^{\frac12}}{[1]\left(\frac{1}{\delta}+1+1\right)^{m-1}} \\
&\quad+\frac{\left(\beta^{\frac12}(\frac{t^2+1}{\delta})+\overline{\beta}^{\frac12}\right)^m\beta^{\frac12}+\left(\overline{\beta}^{\frac12}(\frac{t^2+1}{\delta})+\beta^{\frac12}\right)^m\overline{\beta}^{\frac12}}{[1]\left(\frac{t^2}{\delta}+\frac{1}{\delta}+t\right)^{m-1}} \\
&= \frac{\left[\frac{m+1}{2}\right]t^2}{[2]}+\frac{t}{\delta}+\frac{\left(\beta^{\frac12}+\delta\overline{\beta}^{\frac12}\right)^m\beta^{\frac12}+\left(\overline{\beta}^{\frac12}+\delta\beta^{\frac12}\right)^m\overline{\beta}^{\frac12}}{[2]} \\
&\quad+\frac{\left(\beta^{\frac12}(t^2+1)+\delta\overline{\beta}^{\frac12}\right)^m\beta^{\frac12}+\left(\overline{\beta}^{\frac12}(t^2+1)+\delta\beta^{\frac12}\right)^m\overline{\beta}^{\frac12}}{[2](t^2+\delta t+1)^{m-1}} \\
&= \frac{\left[\frac{m+1}{2}\right]t^2}{[2]}+\frac{t}{\delta}+\frac{\left(\overline{\beta}^{\frac32}\right)^m\beta^{\frac12}+\left(\beta^{\frac32}\right)^m\overline{\beta}^{\frac12}}{[2]} \\
&\quad+\frac{\left(\beta^{\frac12}t^2+\overline{\beta}^{\frac32}\right)^m\beta^{\frac12}+\left(\overline{\beta}^{\frac12}t^2+\beta^{\frac32}\right)^m\overline{\beta}^{\frac12}}{[2](\nu(t))^{2m-2}} \\
&= \frac{\left[\frac{3m-1}{2}\right]}{[2]}+\frac{t}{[1]}+\frac{\left[\frac{m+1}{2}\right]t^2}{[2]}+\frac{\left(\beta t^2+\overline{\beta}\right)^m\overline{\beta}^{\frac{m-1}{2}}+\left(\overline{\beta}t^2+\beta\right)^m\beta^{\frac{m-1}{2}}}{[2](\nu(t))^{2m-2}} \\
&\qquad= \frac{\left[\frac{3m-1}{2}\right]}{[2]}+\frac{t}{[1]}+\frac{\left[\frac{m+1}{2}\right]t^2}{[2]}+\frac{T(a(t)^{2m}\overline{\beta}^{\frac{m-1}{2}})}{[2]\nu(t)^{2m-2}}.
\end{aligned}
$$

This carries the left-hand side of Eq. (3.45) as far as we need. Now consider the right-hand side of Eq. (3.45).

$$
\begin{aligned}
& \frac{1}{[2]}\left\{\left[\frac{m+1}{2}\right]\left\{\frac{[m+1]}{[2]}+\frac{[m-1]}{[2]}t^2+\frac{T(a(t)^{2m}\overline{\beta})}{[2]\nu(t)^{2(m-1)}}+\frac{t}{\delta}\right\}+\right. \\
&\quad+\left[\frac{m+1}{4}\right]\left[\frac{m-3}{4}\right]t+\left[\frac{m-3}{2}\right]\left\{\frac{[m-1]}{[2]}+\frac{[m+1]}{[2]}t^2+\frac{T(a(t)^{2m}\beta)}{[2]\nu(t)^{2m-2}}+\frac{t}{\delta}\right\} \\
&= \frac{1}{[2]}\left\{\frac{\left[\frac{m+1}{2}\right]\left[\frac{2m+2}{2}\right]+\left[\frac{m-3}{2}\right]\left[\frac{2m-2}{2}\right]}{[2]}+t\left(\frac{\left[\frac{m+1}{2}\right]}{[1]}+\frac{\left[\frac{m+1}{4}\right]\left[\frac{m-3}{4}\right][1]}{[1]}+\frac{\left[\frac{m-3}{2}\right]}{[1]}\right)\right.
\end{aligned}
$$

$$+\left(\frac{\left[\frac{m+1}{2}\right]\left[\frac{2m-2}{2}\right]+\left[\frac{m-3}{2}\right]\left[\frac{2m+2}{2}\right]}{[2]}\right)t^2$$

$$+\frac{\left[\frac{m+1}{2}\right]T(a(t)^{2m}\overline{\beta})+\left[\frac{m-3}{2}\right]T(a(t)^{2m}\beta)}{[2]\nu(t)^{2m-2}}\Bigg\}$$

$$=\frac{1}{[2]}\Bigg\{\frac{\left[\frac{3m+3}{2}\right]+\left[\frac{m+1}{2}\right]+\left[\frac{3m-5}{2}\right]+\left[\frac{m+1}{2}\right]}{[2]}$$

$$+t\left(\frac{\left[\frac{m+1}{2}\right]+\left(\left[\frac{2m-2}{4}\right]+\left[\frac{4}{4}\right]\right)\left[\frac{4}{4}\right]+\left[\frac{m-3}{2}\right]}{[1]}\right)$$

$$+t^2\left(\frac{\left[\frac{3m-1}{2}\right]+\left[\frac{m-3}{2}\right]+\left[\frac{3m-1}{2}\right]+\left[\frac{m+5}{2}\right]}{[2]}\right)$$

$$+\frac{\left(\beta^{\frac{m+1}{2}}+\overline{\beta}^{\frac{m+1}{2}}\right)T(a(t)^{2m}\overline{\beta})+\left(\beta^{\frac{m-3}{2}}+\overline{\beta}^{\frac{m-3}{2}}\right)T(a(t)^{2m}\beta)}{[2]\nu(t)^{2m-2}}\Bigg\}$$

$$=\frac{1}{[2]}\Bigg\{\frac{\left[\frac{3m-1}{2}\right][2]}{[2]}+t\left(\frac{\left[\frac{m+1}{2}\right]+\left[\frac{2m+2}{4}\right]+\left[\frac{2m-6}{4}\right]+[2]+\left[\frac{m-3}{2}\right]}{[1]}\right)$$

$$+t^2\left(\frac{\left[\frac{m+1}{2}\right]\left[\frac{4}{2}\right]}{[2]}\right)$$

$$+\frac{(\beta^{\frac{m+1}{2}}+\overline{\beta}^{\frac{m+1}{2}})(a(t)^{2m}\overline{\beta}+\overline{a}(t)^{2m}\beta)+(\beta^{\frac{m-3}{2}}+\overline{\beta}^{\frac{m-3}{2}})(a(t)^{2m}\beta+\overline{a}(t)^{2m}\overline{\beta}}{[2]\nu(t)^{2m-2}}\Bigg\}$$

$$=\frac{1}{[2]}\Bigg\{\left[\frac{3m-1}{2}\right]+t[1]+t^2\left[\frac{m+1}{2}\right]+\frac{a(t)^{2m}\left(\beta^{\frac{m-1}{2}}+\overline{\beta}^{\frac{m+3}{2}}+\beta^{\frac{m-1}{2}}+\overline{\beta}^{\frac{m-5}{2}}\right)}{[2]v(t)^{2m-2}}$$

$$+\frac{\overline{a}(t)^{2m}\left(\beta^{\frac{m+3}{2}}+\overline{\beta}^{\frac{m-1}{2}}+\beta^{\frac{m-5}{2}}+\overline{\beta}^{\frac{m-1}{2}}\right)}{[2]\nu(t)^{2m-2}}\Bigg\}$$

$$=\frac{\left[\frac{3m-1}{2}\right]}{[2]}+\frac{t}{[1]}+t^2\frac{\left[\frac{m+1}{2}\right]}{[2]}$$

$$+\frac{a(t)^{2m}\left(\overline{\beta}^{\frac{m+3}{2}}+\overline{\beta}^{\frac{m-5}{2}}\right)+\overline{a}(t)^{2m}\left(\beta^{\frac{m+3}{2}}+\beta^{\frac{m-5}{2}}\right)}{[2]^2v(t)^{2m-2}}$$

$$= \frac{\left[\frac{3m-1}{2}\right]}{[2]} + \frac{t}{[1]} + t^2 \frac{\left[\frac{m+1}{2}\right]}{[2]} + \frac{a(t)^{2m}\overline{\beta}^{\frac{m-1}{2}}\left(\overline{\beta}^{\frac{4}{2}} + \overline{\beta}^{-\frac{4}{2}}\right) + \overline{a}(t)^{2m}\beta^{\frac{m-1}{2}}\left(\beta^{\frac{4}{2}} + \beta^{-\frac{4}{2}}\right)}{[4]v(t)^{2m-2}}$$

$$= \frac{\left[\frac{3m-1}{2}\right]}{[2]} + \frac{t}{[1]} + t^2 \frac{\left[\frac{m+1}{2}\right]}{[2]} + \frac{T(a(t)^{2m}\overline{\beta}^{\frac{m-1}{2}})}{[2]v(t)^{2m-2}}.$$

This is the same formula we obtained earlier for the left-hand side of Eq. (3.45), completing the proof. □

At this point we know that each of the four infinite families of cyclic q-clans satisfies the hypotheses of Theorem 3.10.7. This means that the corresponding $GQ(\mathcal{C})$ admits a collineation group isomorphic to

$$\tilde{G} = \langle \tau_m, \psi \rangle \cong C_{q+1} \rtimes C_{2e}$$

fixing the points $(0,0,0)$ and (∞). Moreover, as this is a maximal subgroup of $P\Gamma L(2,q)$ (see [Di58]), the group $\hat{G}_0$ is either just $\tilde{G}$ or is all of $P\Gamma L(2,q)$. Hence, if $\mathcal{C}$ is non-classical we can say the same thing about the group G_0. We will determine the complete oval stabilizers in the Subiaco and Adelaide cases and thus get an easy proof that G_0 is completely determined. A similar approach will be used for the Payne examples.

4.8 An Oval Stabilizer

In this section we will adopt the normalization according to Observation 1.6.6. For the cyclic GQ there is only one case where the group G_0 is not transitive on the ovals of the herd. This is the case in the Subiaco GQ when $q = 2^e$ with $e \equiv 2 \pmod 4$, and in this case there are actually two inequivalent ovals. But in all cases the oval stabilizer studied below does occur.

Put

$$A = \begin{pmatrix} 1 & \delta^{-\frac{1}{2}} \\ 0 & \delta^{-\frac{1}{2}} \end{pmatrix} \text{ and } B = \frac{1}{\left[\frac{1}{4}\right]^3} \begin{pmatrix} \left[\frac{m-3}{4}\right] & \left[\frac{m-1}{4}\right] \\ \left[\frac{m+1}{4}\right] & \left[\frac{m+3}{4}\right] \end{pmatrix}.$$

At this point (see Theorem 3.10.7) we know that $\theta = \theta(2, A \otimes B)$ generates a cyclic group of order $2e$ stabilizing the oval $\mathcal{O}_\alpha$ where $\alpha = \left(\left[\frac{m+1}{4}\right], \left[\frac{m-1}{4}\right]\right)$. It is, in fact, an easy exercise to check that $\alpha^{(2)}B = \lambda\alpha$, where $\lambda = \left[\frac{1}{4}\right]$.

Use Theorem 2.8.1 (dividing the right-hand image by λ^2) to determine the induced oval stabilizer

$$\hat{\theta} : (1, t, g(\alpha, t)) \mapsto ((1, t^2)A^{(2)}, \frac{\delta^{-\frac{1}{2}}}{\delta^{\frac{1}{2}}}(g(\alpha, t))^{(2)} + 1 \cdot g(\alpha, \frac{1}{\delta}) + 0)$$

$$= (1, \frac{1+t^2}{\delta}, \frac{1}{\delta}(\alpha A_t \alpha^T)^{(2)} + \alpha A_{(\frac{1}{\delta})}\alpha^T).$$

At this point put $t = \frac{[j+1]}{[j]}$, so $\frac{1+t^2}{\delta} = \frac{[2j+1]}{[2j]}$, and $\frac{1}{\delta} = \frac{[-2+1]}{[-2]}$. Then compute

$$\alpha A_t \alpha^T = \left(\left[\frac{m+1}{4}\right], \left[\frac{m-1}{4}\right] \right) \begin{pmatrix} f(t) & t^{\frac{1}{2}} \\ 0 & g(t) \end{pmatrix} \begin{pmatrix} \left[\frac{m+1}{4}\right] \\ \left[\frac{m-1}{4}\right] \end{pmatrix}$$
$$= \left[\frac{m+1}{2}\right] f(t) + \left[\frac{m+1}{4}\right]\left[\frac{m-1}{4}\right] t^{\frac{1}{2}} + \left[\frac{m-1}{2}\right] g(t)$$
$$= \left[\frac{m+1}{2}\right] \left\{ \frac{\left[\frac{(j+1)(m-1)}{2}\right]\left[\frac{j(m+1)}{2}\right]}{[j]} + \left(\frac{[j+1]}{[j][1]}\right)^{\frac{1}{2}} \right\} + \frac{\left[\frac{m+1}{4}\right]\left[\frac{m-1}{4}\right]\left[\frac{j+1}{2}\right]}{\left[\frac{j}{2}\right]}$$
$$+ \left[\frac{m-1}{2}\right] \left\{ \frac{\left[\frac{(j+1)(m+1)}{2}\right]\left[\frac{j(m-1)}{2}\right]}{[j]} + \left(\frac{[j+1]}{[j][1]}\right)^{\frac{1}{2}} \right\}$$
$$= \frac{[jm][1] + [m][j] + [j+1] + [j] + [1]}{[j]},$$

after a couple routine steps. So with $t = \frac{[j+1]}{[j]}$, the preceding equality says that

$$g(\alpha, t) = \frac{[jm][1] + [m][j] + [j+1] + [j] + [1]}{[j]}. \tag{4.31}$$

Then with $t = \frac{1}{\delta} = \frac{[-2+1]}{[-2]}$, we have:

$$g(\alpha, \frac{1}{\delta}) = \frac{[2m] + [m][1] + [1]}{[1]}. \tag{4.32}$$

After a little simplification we finally see that

$$\hat{\theta} : (1, t, g(\alpha, t)) \mapsto (1, t, g(\alpha, t))^{(2)} \begin{pmatrix} 1 & \frac{1}{\delta} & g(\alpha, \frac{1}{\delta}) \\ 0 & \frac{1}{\delta} & 0 \\ 0 & 0 & \frac{1}{\delta} \end{pmatrix},$$

which turns out to be the same as

$$\hat{\theta} : (1, t, g(\alpha, t)) \mapsto \left(1, \frac{[2j+1]}{[2j]}, \frac{[2jm][1] + [m][2j] + [2j+1] + [2j] + [1]}{[2j]}\right). \tag{4.33}$$

If we put

$$p_j = \left([j], [j+1], [jm][1] + [m][j] + \left[\frac{j+1}{2}\right]\left[\frac{j}{2}\right]\left[\frac{1}{2}\right] \right),$$

for j modulo $q+1$, then

$$\mathcal{O}_\alpha = \{p_j : j \pmod{q+1}\}$$

is the oval and $\hat{\theta} : p_j \mapsto p_{2j}$. Since $2^e = q \equiv -1 \pmod{q+1}$, $\hat{\theta}^e : p_j \mapsto p_{-j} = ([j], [j-1], [jm][1] + [m][j] + \left[\frac{j-1}{2}\right]\left[\frac{j}{2}\right]\left[\frac{1}{2}\right])$.

Put

$$A = \begin{pmatrix} 1 & \frac{1}{\delta} & g(\alpha, (\frac{1}{\delta})) \\ 0 & \frac{1}{\delta} & 0 \\ 0 & 0 & \frac{1}{\delta} \end{pmatrix}$$

and

$$B = \begin{pmatrix} 1 & 0 & \frac{[m]+1}{\delta} \\ 0 & 1 & \delta^{-1} \\ 0 & 0 & \delta^{-1} \end{pmatrix}, \quad \text{so } B^{-1} = \begin{pmatrix} 1 & 0 & [m]+1 \\ 0 & 1 & 1 \\ 0 & 0 & \delta \end{pmatrix}.$$

Use the linear map $(x, y, z) \mapsto (x, y, z)B$ to replace the oval $\mathcal{O}_\alpha$ with the oval

$$\mathcal{O}'_\alpha = \left\{p'_j = ([j], [j+1], [jm] + 1) : j \pmod{q+1}\right\}.$$

Then $\hat{\theta}$ induces the map $\hat{\theta}'$ on $\mathcal{O}'_\alpha$ given by

$$(x, y, z) \mapsto (x, y, z)^{(2)}(B^{-1})^{(2)}AB = (x^2, y^2, z^2)\begin{pmatrix} 1 & \delta^{-1} & 0 \\ 0 & \delta^{-1} & 0 \\ 0 & 0 & 1 \end{pmatrix}.$$

Up to this point there are no general results putting restrictions on those m for which the cyclic construction actually works, although for small fields the four examples known are indeed the only ones. However, since for the m that give q-clans we also have the oval $\mathcal{O}'_\alpha$, it would be interesting to see if there is an easy way to check for which m the set $\mathcal{O}'_\alpha$ is an oval. This is if and only if for distinct a, b and c mod $j+1$ it is always true that

$$0 \neq \begin{vmatrix} [a] & [a+1] & [am]+1 \\ [b] & [b+1] & [bm]+1 \\ [c] & [c+1] & [cm]+1 \end{vmatrix},$$

which is if and only if

$$[am][b-c] + [bm][c-a] + [cm][a-b] \neq \left[\frac{a-b}{2}\right]\left[\frac{b-c}{2}\right]\left[\frac{c-a}{2}\right]. \tag{4.34}$$

This is as far as we have progressed on this problem (except that we can use the above to show that m cannot be equal to $\frac{1}{4}$).

Chapter 5

Applications to the Known Cyclic q-Clans

We will use the normalization according to Observation 1.6.6.

5.1 The Classical Examples: $\mathrm{q} = 2^{\mathrm{e}}$ for $\mathrm{e} \geq 1$

To obtain the canonical form of the classical q-clan put $k = m = 1$ in Eqs. (4.7) and (4.8). A simple computation shows that if $w = \frac{1}{\delta^{\frac{1}{2}}}$, so $tr(w) = 1$, then the classical q-clan $\mathcal{C}$ in $\frac{1}{2}$-normalized form is given by

$$\mathcal{C} = \left\{ A_t = \begin{pmatrix} wt^{\frac{1}{2}} & t^{\frac{1}{2}} \\ 0 & wt^{\frac{1}{2}} \end{pmatrix} : t \in F \right\}. \tag{5.1}$$

It is an easy exercise to show that the flock $\mathcal{F}(\mathcal{C})$ is *linear*, i.e., all the planes of the flock contain a common line. It requires considerably more work to show that $GQ(\mathcal{C})$ is isomorphic to the classical GQ $H(3, q^2)$, which is the point-line dual of $Q(5, q)$. The notation is a bit different, but the general proof is in 10.6.1 of [PT84].

All the computations below are quite well known, but we find it convenient to include them here.

It is well known that the orthogonal group $PGO^-(6, q)$ is transitive on the set of incident point-line pairs of $Q(5, q)$, and that $|PGO_-(6, q)| = 2q^6(q^4 - 1) \cdot (q^3 + 1)(q^2 - 1)$ (cf. [Hi98], p. 420). We are now going to use the Fundamental Theorem to show that the collineation group $\mathcal{G}_0$ of $GQ(\mathcal{C})$ fixing (∞) and $(0, 0, 0)$ has the correct order, i.e., the order that it should have if it is really true that $GQ(\mathcal{C}) \cong H(3, q^2)$. First note that $GQ(\mathcal{C})$ has $(1 + q^2)(1 + q^3)$ points, and that the number of points not collinear with (∞) is q^5.

We will show that $|\mathcal{G}_0| = 2q(q^2-1)^2e$, where $q = 2^e$. An outline of the proof is as follows. First we establish that there is a special linear subgroup of $\mathcal{G}_0$ acting sharply triply transitively on the lines through (∞). This subgroup contributes $q(q^2-1)$ to the order of $\mathcal{G}_0$. Then we show that for each $\sigma \in Aut(F)$ there are exactly $2(q+1)$ special collineations with companion automorphism equal to σ and which fix the three lines $[A(\infty)]$, $[A(0)]$ and $[A(1)]$ through (∞).

First, by Theorem 4.4.1 we see that $\theta(id, P \otimes P)$ is a collineation of $GQ(\mathcal{C})$ interchanging $[A(\infty)]$ and $[A(0)]$.

Second, since $g(\alpha,t) = \alpha A_t \alpha^T = t^{\frac{1}{2}}\left(\alpha \left(\begin{smallmatrix} w & 1 \\ 0 & w \end{smallmatrix}\right) \alpha^T\right)$, the map $t \mapsto g_t$ is additive. For $r \in F$, consider the automorphism of $G^{\otimes}$ defined by:

$$\theta = \theta(id, \begin{pmatrix} 1 & r^{\frac{1}{2}} \\ 0 & 1 \end{pmatrix} \otimes I) : (\alpha, \beta, c) \mapsto$$

$$((\alpha,\beta)[\begin{pmatrix} 1 & r^{\frac{1}{2}} \\ 0 & 1 \end{pmatrix} \otimes I], c + (\alpha,\beta) \begin{pmatrix} A_r & 0 \\ 0 & 0 \end{pmatrix} (\alpha,\beta)^T)$$
$$= (\alpha, r^{\frac{1}{2}}\alpha + \beta, c + g(\alpha, r)).$$

Here

$$\theta : (\gamma_{yt} \otimes \alpha, g(\alpha,t)) \mapsto (\gamma_{yt} \begin{pmatrix} 1 & r^{\frac{1}{2}} \\ 0 & 1 \end{pmatrix} \otimes \alpha, g(\alpha,t) + g(\alpha,r)) = (\gamma_{yt+r} \otimes \alpha, g(\alpha, t+r)).$$

Hence $\theta : A(t) \to A(t+r)$ for all $t \in \tilde{F}$, so $\theta \in \mathcal{G}_0$. These two results show that $\mathcal{G}_0$ is doubly transitive on the lines through (∞).

Third, for $0 \neq \lambda \in F$, put $\bar{t} = \lambda^2 t$ (so $\bar{0} = 0$). Then $A_{\bar{t}} = \bar{t}^{\frac{1}{2}} \left(\begin{smallmatrix} w & 1 \\ 0 & w \end{smallmatrix}\right) = \lambda t^{\frac{1}{2}} \left(\begin{smallmatrix} w & 1 \\ 0 & w \end{smallmatrix}\right) = \lambda I^{-1}(t^{\frac{1}{2}} \left(\begin{smallmatrix} w & 1 \\ 0 & w \end{smallmatrix}\right)) I^{-T} + A_0$. So by the Fundamental Theorem $\theta(id, \left(\begin{smallmatrix} 1 & 0 \\ 0 & \lambda \end{smallmatrix}\right) \otimes I) : (\alpha, \beta, c) \mapsto (\alpha, \lambda\beta, \lambda c)$ is an element of $\mathcal{G}_0$ mapping $[A(t)]$ to $[A(\lambda^2 t)]$, so fixing $[A(\infty)]$ and $[A(0)]$. Hence $\mathcal{G}_0$ is triply transitive on the lines through (∞).

Moreover, the collineations that we have just established to show that $\mathcal{G}_0$ is triply transitive on the lines through (∞) are all special linear. So by Theorem 2.3.10 and the Note following it we see that this special linear group is *sharply* triply transitive on these lines.

For the classical q-clan $\mathcal{C}$ (as for *all* q-clans), there is the kernel $\mathcal{N} = \{\theta(id, I \otimes aI) : (\alpha, \beta, c) \mapsto (a\alpha, a\beta, a^2c) : 0 \neq a \in F\}$. This gives a group of order $q-1$ fixing all lines through (∞). However, since we are in the classical case, i.e., the linear case, we cannot invoke Theorem 2.5.4. But we do know that to determine $|\mathcal{G}_0|$ it suffices to determine all collineations of the form $\theta(\sigma, I \otimes B) : (\alpha, \beta, c) \mapsto ((\alpha^{(\sigma)}, \beta^{(\sigma)})(I \otimes B), c^{\sigma}) = (\alpha^{(\sigma)}B, \beta^{(\sigma)}B, c^{\sigma})$.

Our next step is to show that for each $\sigma \in Aut(F)$ there is a collineation θ with companion automorphism σ and fixing $[A(\infty)]$, $[A(0)]$, and $[A(1)]$, i.e., $\theta = \theta(\sigma, I \otimes B)$. To this end, for each $a \in F$, let $M(a) = \begin{pmatrix} a+1 & a \\ a & a+1 \end{pmatrix}$.

Then $M(a)M(b) = M(a+b)$. So $a \mapsto M(a)$ is an isomorphism from the additive group of F to the multiplicative group of matrices of the form $M(a)$. Note that $M(0) = I$, $M(1) = P$, and for $\sigma \in Aut(F)$, $(M(a))^{(\sigma)} = M(a^\sigma)$. Now consider the automorphism θ of $G^\otimes$ given by $\theta = \theta(2, I \otimes M(w))$. $\theta^2 = \theta(2^2, I \otimes M(w^2)M(w)) = \theta(2^2, I \otimes M(w^2 + w))$. By induction it follows that $\theta^e = \theta(2^e, I \otimes M(tr(w))) = \theta(id, I \otimes P)$. Hence the order of θ is $2e$. It is also readily checked that $M(w)^{-1}A_t^{(2)}M(w)^{-T} \equiv A_{t^2}$. Hence by Theorem 2.3.10, $\theta = \theta(2, I \otimes M(w))$ is a collineation of $GQ(\mathcal{C})$ mapping $[A(t)]$ to $[A(t^2)]$. And θ^i has companion automorphism equal to 2^i.

Note that since $(1,1)^{(2)}M(w) = (1,1)$, it also follows that θ generates a group of order $2e$ leaving invariant $\bar{\mathcal{O}}_{(1,1)}$.

For each $\sigma \in Aut(F)$, let $\theta(\sigma, I \otimes B_\sigma)$ be a particular special collineation fixing the three lines $[A(\infty)]$, $[A(0)]$, and $[A(1)]$ and having companion automorphism σ. Without loss of generality and for convenience a little later, we may suppose that $\theta(\sigma, I \otimes B_\sigma)$ fixes the oval $\bar{\mathcal{O}}_{(1,1)}$. (In fact, we could assume that $\theta(\sigma, I \otimes B_\sigma) = \theta^i$ if $\sigma = 2^i$.)

Now suppose that $\theta(\sigma, I \otimes B)$ is *any* special collineation fixing the three lines $[A(\infty)]$, $[A(0)]$, and $[A(1)]$ and having companion automorphism σ. Then $\theta(\sigma, I \otimes B) = \theta(\sigma, I \otimes B_\sigma) \circ \theta(id, I \otimes B_\sigma^{-1}B)$, where $B_\sigma^{-1}B = D$ is completely determined by $det(D) = 1$, and $A_t \equiv D^{-1}A_tD^{-T}$. So to complete the computation of $|\mathcal{G}_0|$ we need only count the number k of such matrices D. Then $|\mathcal{G}_0| = k(q+1)q(q-1)^2e$.

So we need to determine all $D^{-1} = \begin{pmatrix} a & b \\ c & d \end{pmatrix}$ satisfying

$$\begin{array}{rl} \text{(i)} & a^2w + ab + b^2w = w; \\ \text{(ii)} & ad + bc = 1; \\ \text{(iii)} & c^2w + cd + d^2w = w. \end{array} \tag{5.2}$$

If $b = 0$, it is easy to show that the only possibilities are $D^{-1} = I$ or $D^{-1} = \begin{pmatrix} 1 & 0 \\ w^{-1} & 1 \end{pmatrix} = D$. Suppose $b \neq 0$. Then (i) has a solution for a if and only if $tr\left(\frac{w(b^2w+w)}{b^2}\right) = tr(w^2 + (w/b)^2) = 0$ if and only if $tr(w/b) = 1$. There are $q/2$ solutions for such b, and for each such b, two solutions for a. So there are q pairs (a,b) with $a^2w + ab + b^2w = w$, and $b \neq 0$. From (ii) put $c = \frac{1+ad}{b}$. Then (iii) is equivalent to $\left(\frac{1+ad}{b}\right)^2 w + \left(\frac{1+ad}{b}\right) d + d^2w = w$, which is equivalent to $w + d^2(a^2w+ab+b^2w)+bd+b^2w = 0$, which holds if and only if $d^2w+db+b^2w+w = 0$, which has exactly two solutions for d since $tr\left(\frac{w(b^2w+w)}{b^2}\right) = tr\left(w^2 + \left(\frac{w}{b}\right)^2\right) = 0$. So for each of the q pairs (a,b) with $b \neq 0$, there are two pairs (c,d) satisfying (i), (ii) and (iii). So if we include the two solutions for $b = 0$, we have $2(q+1)$

solutions for D^{-1} such that $D^{-1}A_tD^{-T} \equiv A_t$ for all $t \in F$. Hence

$$|\mathcal{G}_0| = 2(q+1)[(q+1)q(q-1)](q-1)e = 2q(q^2-1)^2e,$$

as claimed.

The oval stabilizer

We now apply to the classical examples the theory developed concerning the stabilizers of the associated ovals.

Put $\alpha = (1,1)$ and consider the oval $\mathcal{O}_\alpha = \{(\gamma_t, g(\alpha,t)) : t \in \tilde{F}\} = \{(1,t,t^{\frac{1}{2}}) : t \in F\} \cup \{(0,1,0)\} = \{(x_0,x_1,x_2) \in PG(2,q) : x_0x_1 = x_2^2\}$. Earlier we found the following elements of $\mathcal{G}_0$: $\phi = \theta(id, P \otimes P) \in \mathcal{G}_0$, and $(1,1)P = (1,1)$, so ϕ fixes $\mathcal{R}_{(1,1)}$. The corresponding $\hat{\phi}$ acts on $\mathcal{O}_\alpha$ as: $\hat{\phi} = \hat{\theta}(id, P\otimes P) : (\gamma_t, g(\alpha,t)) \mapsto (\gamma_t P, g(\alpha,t))$, i.e., $\hat{\phi}(id, P\otimes P) : (1,t,t^{\frac{1}{2}}) \mapsto (t,1,t^{\frac{1}{2}})$. This says $\hat{\phi}$ is the collineation

$$\hat{\phi} : (x,y,z) \mapsto (x,y,z)\begin{pmatrix} 0 & 1 & 0 \\ 1 & 0 & 0 \\ 0 & 0 & 1 \end{pmatrix} = (y,x,z).$$

Second, for any $r \in F$, $\theta(id, \begin{pmatrix} 1 & r^{\frac{1}{2}} \\ 0 & 1 \end{pmatrix} \otimes I) \in \mathcal{G}_0$ and $(1,1)I = (1,1)$. Here $C = a_4^2 A_{(a_2/a_4)^2} = A_r = r^{\frac{1}{2}}\left(\begin{smallmatrix} r & 1 \\ 0 & w \end{smallmatrix}\right)$, and $E = 0$ (recall Eq. (2.26)). So

$$\hat{\theta}(id, \begin{pmatrix} 1 & r^{\frac{1}{2}} \\ 0 & 1 \end{pmatrix} \otimes I) : (\gamma_t, g(\alpha,t)) \mapsto (\gamma_t \begin{pmatrix} 1 & r^{\frac{1}{2}} \\ 0 & 1 \end{pmatrix}^{(2)}, g(\alpha,t) + \gamma_t(r^{\frac{1}{2}}, 0)^T),$$

i.e.,

$$\hat{\theta}(id, \begin{pmatrix} 1 & r^{\frac{1}{2}} \\ 0 & 1 \end{pmatrix} \otimes I) : (1,t,t^{\frac{1}{2}}) \mapsto ((1,t)\begin{pmatrix} 1 & r \\ 0 & 1 \end{pmatrix}, t^{\frac{1}{2}} + r^{\frac{1}{2}}) = (1, t+r, (t+r)^{\frac{1}{2}}).$$

Third, for $0 \neq \lambda \in F$, $\theta(id, \left(\begin{smallmatrix} 1 & 0 \\ 0 & \lambda \end{smallmatrix}\right) \otimes I) \in \mathcal{G}_0$ and $(1,1)I = (1,1)$. Here $C = A_0 = 0$, $E = 0$. So $\hat{\theta}(id, \left(\begin{smallmatrix} 1 & 0 \\ 0 & \lambda \end{smallmatrix}\right) \otimes I) : (1,t,t^{\frac{1}{2}}) \mapsto ((1,t)\left(\begin{smallmatrix} 1 & 0 \\ 0 & \lambda^2 \end{smallmatrix}\right), \lambda t^{\frac{1}{2}}) = (1, \lambda^2 t, \lambda t^{\frac{1}{2}})$.

At this point we have the following three types of collineations of $PG(2,q)$ that generate a linear group stabilizing $\mathcal{O}_{(1,1)}$ and whose inducing collineations of $GQ(\mathcal{C})$ act sharply triply transitively on the lines through (∞).

$$\begin{aligned} \hat{\phi} = \hat{\theta}(id, P\otimes P) : &\quad (x,y,z) \mapsto (y,x,z); \\ \hat{\theta}(id, \begin{pmatrix} 1 & r^{\frac{1}{2}} \\ 0 & 1 \end{pmatrix} \otimes I) : &\quad (x_0,x_1,x_2) \mapsto (x_0, rx_0 + x_1, r^{\frac{1}{2}}x_0 + x_2); \\ \hat{\theta}(id, \begin{pmatrix} 1 & 0 \\ 0 & \lambda \end{pmatrix} \otimes I) : &\quad (x_0,x_1,x_2) \mapsto (x_0, \lambda^2 x_1, \lambda x_2). \end{aligned} \tag{5.3}$$

Here $r \in F$ and $0 \neq \lambda \in F$.

Also earlier in this section it was determined that $\theta(2, I \otimes M(w))$ generates a group of order $2e$ of collineations fixing each of the lines $[A(\infty)]$, $[A(0)]$, $[A(1)]$ and leaving invariant $\overline{\mathcal{O}}_{(1,1)}$. Here $C = D = E = 0$, so that $\hat{\theta}(2, I \otimes M(w))\colon (1, t, t^{\frac{1}{2}}) \mapsto ((1, t^2)I^{(2)}, (t^{\frac{1}{2}})^2) = (1, t^2, t)$. So

$$\hat{\theta}(2, I \otimes M(w)) : (x_0, x_1, x_2) \mapsto (x_0^2, x_1^2, x_2^2). \tag{5.4}$$

It follows that the subgroup of $\mathcal{G}_0$ stabilizing the conic $\overline{\mathcal{O}}_{(1,1)}$ induces the group of collineations of $PG(2, q)$ stabilizing the conic $\mathcal{O}_{(1,1)}$, which is known to be the complete stabilizer in $P\Gamma O(3, q)$ of the conic.

Finally, it is easy to check that each oval $\mathcal{O}_\alpha$ is a conic, and we want to see that the group $\mathcal{G}_0$ is transitive on these conics. Recall that we produced a special linear group of order $q(q^2-1)$ acting sharply triply transitively on the lines through (∞) and fixing $\overline{\mathcal{O}}_{(1,1)}$. Then there was a disjoint cyclic group of order $2e$ fixing the lines $[A(\infty)]$, $[A(0)]$ and $[A(1)]$, and fixing $\bar{\mathcal{O}}_{(1,1)}$. Also there is the kernel $\mathcal{N}$ of order $q - 1$ fixing all lines through (∞) and fixing $\bar{\mathcal{O}}_{(1,1)}$. So in the group $\mathcal{G}_0$ with order $2q(q^2 - 1)^2 e$, we have a stabilizer of $\overline{\mathcal{O}}_{(1,1)}$ having order $q(q^2 - 1)2e(q - 1)$. If we could show that this is the complete stabilizer of $\bar{\mathcal{O}}_{(1,1)}$, then the $\mathcal{G}_0$-orbit on the ovals that contains $\bar{\mathcal{O}}_{(1,1)}$ would have order $q + 1$.

Just for this proof let $\mathcal{T}$ be the subgroup of $\mathcal{G}_0$ stabilizing $\bar{\mathcal{O}}_{(1,1)}$. This group also acts on the lines through (∞) and is triply transitive on them. Modulo the kernel $\mathcal{N}$, each collineation of $\mathcal{T}$ fixing the three lines $[A(\infty)]$, $[A(0)]$ and $[A(1)]$, must have the form

$$\theta(\sigma, I \otimes B) = \theta(\sigma, I \otimes B_\sigma) \circ \theta(id, I \otimes B_\sigma^{-1} B),$$

where B_σ is the matrix considered above, and where $B_\sigma^{-1} B = D$ not only satisfies $det(D) = 1$, and $A_t \equiv D^{-1} A_t D^{-T}$, but also it must fix $\bar{\mathcal{O}}_{(1,1)}$. So in addition to the conditions placed on D^{-1} in Eq. (5.2) we must also have the following condition satisfied:

$$(iv) \quad a + b + c + d = 0 \tag{5.5}$$

Now using the conditions (i) through (iv) of Eqs. (5.2) and (5.5) it is possible to show that one must have either $D^{-1} = D = P$ or $D^{-1} = D = I$. It follows that if $\sigma = 2^i$, then either $\theta(\sigma, I \otimes B) = \theta^i$ or $\theta(\sigma, I \otimes B) = \theta^{i+e}$, where $\theta = \theta(2, I \otimes M(w))$. This shows that $\mathcal{T}$ has order $q(q^2 - 1)2e(q - 1)$ as desired.

5.2 The FTWKB Examples: $\mathbf{q = 2^e}$ with e Odd

Here $m = -\frac{q}{2} \equiv \frac{1}{2}(mod q + 1)$. If we compute the *canonical form* of the q-clan we get

$$f(t) = t^{\frac{1}{4}}(1 + \delta^{-\frac{1}{2}}) + t^{\frac{1}{2}}(\delta^{-\frac{1}{4}} + \delta^{-\frac{1}{2}}) + \delta^{-\frac{1}{2}} t^{\frac{3}{4}},$$

and

$$g(t) = t^{\frac{1}{4}}(\delta^{-\frac{1}{2}}) + t^{\frac{1}{2}}(\delta^{-\frac{1}{4}} + \delta^{-\frac{1}{2}}) + t^{\frac{3}{4}}(1 + \delta^{-\frac{1}{2}}).$$

Put $Q = \begin{pmatrix} 1+\delta^{-\frac{1}{4}} & \delta^{-\frac{1}{4}} \\ \delta^{-\frac{1}{4}} & 1+\delta^{-\frac{1}{4}} \end{pmatrix}$. Then $Q = Q^{-1} = Q^T$ and

$$Q \begin{pmatrix} f(t) & t^{\frac{1}{2}} \\ 0 & g(t) \end{pmatrix} Q \equiv \begin{pmatrix} t^{\frac{1}{4}} & t^{\frac{1}{2}} \\ 0 & t^{\frac{3}{4}} \end{pmatrix}.$$

The form on the right is the *standard form* of the matrices of the q-clan.

It is possible to discuss this case abstractly with almost no computations, since we already know that G_0 ($\overline{G}_0$) contains a subgroup isomorphic to $C_{q+1} \rtimes C_{2e}$ that contains only two semi-linear elements with associated field automorphism $\sigma = 2$. In this case

$$A_{t^\sigma} \equiv A_t^\sigma,$$

so that the generator ρ_σ corresponding to the collineation $\theta(\sigma, I \otimes I)$ belongs to G_0 for any $\sigma \in Aut(F)$. This forces the group to be larger than $C_{q+1} \rtimes C_{2e}$ and hence to be the entire group $P\Gamma L(2,q)$. (See below for the details.)

It is an interesting (at least to us!) exercise, however, to show that the generators of $P\Gamma L(2,q)$ under the magic action all belong to $\hat{G}_0$. It gives one confidence that all the general computations we have done are correct. So we illustrate Theorem 3.6.1 with one of the generators (using the notation of Section 3.4).

Let $0 \neq \lambda \in F$. The computations in [OP02] (easy) show that $\sigma_{\lambda 2} : x \mapsto \left(\begin{smallmatrix} \lambda^2 & 0 \\ 0 & 1 \end{smallmatrix}\right) x$ and $\sigma_{\lambda 2} : f_s \mapsto \sigma_{\lambda 2} f_s \in \langle f_{\lambda s} \rangle$. So $B = \left(\begin{smallmatrix} 1 & 0 \\ 0 & \lambda \end{smallmatrix}\right)$. If we adjust B by a scalar (the kernel again) to have determinant 1, we may take $B = \begin{pmatrix} \lambda^{-\frac{1}{2}} & 0 \\ 0 & \lambda^{\frac{1}{2}} \end{pmatrix}$. With $A = \left(\begin{smallmatrix} \lambda^2 & 0 \\ 0 & 1 \end{smallmatrix}\right)$, we have

$$B' = (B^T)^{(\frac{1}{2})} = \begin{pmatrix} \lambda^{-\frac{1}{4}} & 0 \\ 0 & \lambda^{\frac{1}{4}} \end{pmatrix} \text{ and } A' = (A^T)^{(\frac{1}{2})} = \begin{pmatrix} \lambda & 0 \\ 0 & 1 \end{pmatrix}.$$

So $\theta\left(id, \begin{pmatrix} \lambda & 0 \\ 0 & 1 \end{pmatrix} \otimes \begin{pmatrix} \lambda^{-\frac{1}{4}} & 0 \\ 0 & \lambda^{\frac{1}{4}} \end{pmatrix}\right)$ should be in $Aut(GQ(\mathcal{C}))$. To put this in the form given in the Fundamental Theorem we need to adjust A' by a scalar multiple so that $A' = \left(\begin{smallmatrix} 1 & 0 \\ 0 & \lambda^{-1} \end{smallmatrix}\right)$. Then routine computations show that

$$\lambda^{-1} \begin{pmatrix} \lambda^{-\frac{1}{4}} & 0 \\ 0 & \lambda^{\frac{1}{4}} \end{pmatrix}^{-1} \begin{pmatrix} t^{\frac{1}{4}} & t^{\frac{1}{2}} \\ 0 & t^{\frac{3}{4}} \end{pmatrix} \begin{pmatrix} \lambda^{-\frac{1}{4}} & 0 \\ 0 & \lambda^{\frac{1}{4}} \end{pmatrix}^{-T} = \begin{pmatrix} (\lambda^{-2}t)^{\frac{1}{4}} & (\lambda^{-2}t)^{\frac{1}{2}} \\ 0 & (\lambda^{-2}t)^{\frac{3}{4}} \end{pmatrix}.$$

Hence by The Fundamental Theorem the proposed automorphism of $G^{\otimes}$ is indeed a collineation of $GQ(\mathcal{C})$.

Here are the details for the first argument given above. We know that if $C_{q+1} \rtimes C_{2e}$ is the largest subgroup of $P\Gamma L(2,q)$ acting on the herd of ovals, then for the canonical form, $\theta = \theta(2, A \otimes B)$ maps $[A(0)]$ to $[A(\bar{0})]$ where either $\bar{0} = \delta^{-1}$ or $\bar{0} = \delta^{-1} + \delta$, if θ fixes $[A(\infty)]$. In standard form $A_{t\sigma} = A_t^{\sigma} = 1 \cdot I^{-1} A_t^{\sigma} I^{-T} + A_0$. This implies

$$QQ^{(2)} A_t^{(2)} Q^{(2)} Q = A_t 2$$

in canonical form. Hence there is a collineation $\theta(2, \begin{pmatrix} 1 & 0 \\ 0 & 1 \end{pmatrix} \otimes Q^{(2)} Q) \in G_0$ fixing $[A(\infty)]$ and $[A(0)]$.

Since $0 = \bar{0}$ in this case, i.e., there are more than just the two semi-linear collineations belonging to $\sigma = 2$, we know that all of $P\Gamma L(2, q)$ acts on the herd.

The oval in this case is a translation oval associated with the map $x \mapsto x^{\frac{1}{4}}$, i.e., not contained in any *regular* hyperoval (conic plus nucleus). When we check the induced oval stabilizer and compare it with the results of [Hi98] we see that the induced oval stabilizer is the complete oval stabilizer. This implies that we have the complete collineation group of the GQ.

The induced oval stabilizer

Let $\alpha = (1, 0)$ so that $g(\alpha, t) = t^{1/4}$. Hence $\mathcal{O}_\alpha = \{(1, t, t^{\frac{1}{4}}) : t \in F\} \cup \{(0, 1, 0)\} = \{(x_0, x_1, x_2) \in PG(2, q) : x_2^4 = x_0^3 x_1\}$. Consider the collineations $\theta(id, \left(\begin{smallmatrix} 1 & s^{1/2} \\ 0 & 1 \end{smallmatrix}\right) \otimes \left(\begin{smallmatrix} 1 & 0 \\ s^{1/4} & 1 \end{smallmatrix}\right))$ for arbitrary $s \in F$. Since $(1, 0) \left(\begin{smallmatrix} 1 & 0 \\ s^{1/4} & 1 \end{smallmatrix}\right) = (1, 0)$, these collineations fix $\mathcal{R}_{(1,0)}$. Here $C = a_4^2 A_{(a2/a4)^2} = A_s = \left(\begin{smallmatrix} s^{1/4} & s^{1/2} \\ 0 & s^{3/4} \end{smallmatrix}\right)$, $E = a_3^2 A_{(a1/a3)^2} = 0$. So the induced oval stabiliser is

$$\hat{\theta} = \hat{\theta}(id, \begin{pmatrix} 1 & s^{1/2} \\ 0 & 1 \end{pmatrix} \otimes \begin{pmatrix} 1 & 0 \\ s^{1/4} & 1 \end{pmatrix}) : (\gamma_t, g(\alpha, t)) \mapsto$$

$$(\gamma_t \begin{pmatrix} 1 & s \\ 0 & 1 \end{pmatrix}, g(\alpha, t) + \gamma_t (s^{1/4}, 0)^T),$$

i.e., $\hat{\theta} : (1, t, t^{1/4}) \mapsto (1, t + s, (t + s)^{1/4})$. So more generally, $\hat{\theta} : (x_0, x_1, x_2) \mapsto (x_0, x_1, x_2) \left(\begin{smallmatrix} 1 & s & s^{1/4} \\ 0 & 1 & 0 \\ 0 & 0 & 1 \end{smallmatrix}\right) = (x_0, s x_0 + x_1, x_0 s^{\frac{1}{4}} + x_2)$.

For $0 \neq \mu \in F$, $\sigma \in Aut(F)$, $\phi = \theta\left(\sigma, \begin{pmatrix} 1 & 0 \\ 0 & \mu \end{pmatrix} \otimes \begin{pmatrix} \mu^{1/4} & 0 \\ 0 & \mu^{-1/4} \end{pmatrix}\right) \in \mathcal{G}_0$. Since $(1, 0)^\sigma \begin{pmatrix} \mu^{1/4} & 0 \\ 0 & \mu^{-1/4} \end{pmatrix} = \mu^{1/4}(1, 0)$, ϕ leaves $\mathcal{R}_{(1,0)}$ invariant. At this point we have that

$$\hat{\phi} : (\gamma_t, g(\alpha, t)) \mapsto (\mu^{1/2} \gamma_{t\sigma} \left(\begin{smallmatrix} 1 & 0 \\ 0 & \mu^2 \end{smallmatrix}\right), \mu t^{\sigma/4} + \mu^{1/2} \gamma_{t\sigma} (\alpha C \alpha^T, \alpha E \alpha^T)^T),$$

where $C = a_4^2 A_{(a2/a4)^2} = 0$, $E = a_3^2 A_{(a1/a3)^2} = 0$.

So $\hat{\phi}$: $(1, t, t^{1/4}) \mapsto (\mu^{1/2}, \mu^{5/2}t^\sigma, \mu t^\sigma/4)$. And more generally,

$$\hat{\phi}\colon (x_0, x_1, x_2) \mapsto (x_0^\sigma, x_1^\sigma, x_2^\sigma)\begin{pmatrix} \mu^{1/2} & 0 & 0 \\ 0 & \mu^{5/2} & 0 \\ 0 & 0 & \mu \end{pmatrix}.$$

Hence the induced group stabilizing $\mathcal{O}_\alpha$ has order at least $q(q-1)e$, with action $(x_0, x_1, x_2) \mapsto (x_0^\sigma, x_1^\sigma, x_2^\sigma)\begin{pmatrix} \mu^{1/2} & \mu^{1/2}s & \mu^{1/2}s^{1/4} \\ 0 & \mu^{5/2} & 0 \\ 0 & 0 & \mu \end{pmatrix}$. Here μ, $s \in F$, with $\mu \neq 0$, By [Hi98] Cor. 8.27 p. 186, this is the complete stabilizer of $\mathcal{O}_{(1,0)}$. Hence, since $|\mathcal{G}_0| = (q+1)q(q-1)e$, the orbit of $\mathcal{R}_\alpha$ under $\mathcal{G}_0$ has size $q+1$, i.e, $\mathcal{G}_0$ is transitive on all the $\mathcal{R}_\alpha$, $\alpha \in PG(1, q)$. This implies that there is only one oval (up to projective equivalence) arising from the original q-clan, and its oval stabilizer is induced by $\mathcal{G}_0$.

5.3 The Subiaco Examples: $\mathbf{q = 2^e,\ e \geq 4}$

Putting $k = 1$ and $m = 5$ in Eqs 4.7 and 4.8 we obtain the canonical q-clan functions (as in the SN):

$$f(t) = \frac{\delta^2 t^4 + \delta^3 t^3 + (\delta^4 + \delta^2)t^2}{t^4 + \delta^2 t^2 + 1} + \left(\frac{t}{\delta}\right)^{\frac{1}{2}},$$

and

$$g(t) = \frac{(\delta^4 + \delta^2)t^3 + \delta^3 t^2 + \delta^2 t}{t^4 + \delta^2 t^2 + 1} + \left(\frac{t}{\delta}\right)^{\frac{1}{2}}.$$

We know that G_0 and $\hat{G}_0$ contain a subgroup isomorphic to $C_{q+1} \rtimes C_{2e}$. This subgroup induces a stabilizer of each element of the herd which we later show is the complete oval stabilizer in each case. However, the proof that we have the complete oval stabilizer is rather involved. Hence we offer a more elementary proof here that $\hat{G}_0$ can be no larger than $C_{q+1} \rtimes C_{2e}$. (The reader might want to check out the proof in [OP02] that if $\hat{G}_0$ is all of $P\Gamma L(2, q)$, then $\mathcal{S}(\mathcal{C})$ is either classical or FTWKB.) So in particular we have

$$|G_0| = 2(q+1)e.$$

Note that the proof in the SN that the group in this case cannot be larger is quite lengthy, depending on many computations. So we offer here our version of the proof left to the reader in [OP02]. O'Keefe and Penttila make a claim that is equivalent to saying that the generator σ_δ is not in $\hat{G}_0$, which is equivalent to saying that there is no collineation of the form

$$\theta = \theta(id, \begin{pmatrix} 1 & 0 \\ 0 & \delta^{-\frac{1}{2}} \end{pmatrix} \otimes B) \in G_0.$$

Suppose there is such a collineation. Considering the effect of σ_δ on functions, this means there is a matrix B (with determinant equal to 1) for which

$$A_{t/\delta} \equiv \delta^{-\frac{1}{2}} B^{-1} A_t B^{-T} \quad \text{for all } t \in F.$$

This is equivalent to the following matrix equation:

$$\begin{pmatrix} f(t/\delta) \\ g(t/\delta) \end{pmatrix} = \delta^{-\frac{1}{2}} \begin{pmatrix} a^2 & b^2 \\ c^2 & d^2 \end{pmatrix} \begin{pmatrix} f(t) \\ g(t) \end{pmatrix} + \left(\frac{t}{\delta}\right)^{\frac{1}{2}} \begin{pmatrix} ab \\ cd \end{pmatrix}. \tag{5.6}$$

Now consider $f(\frac{1}{t}) = f(\frac{\delta}{t}/\delta)$, that is, replace t with $\frac{\delta}{t}$ in Eq. (5.6). Then use $f(\frac{t}{\delta}) = \frac{t}{\delta} g(\frac{\delta}{t})$ a few times; remember that $ad + bc = 1$, and rearrange the resulting equations in the form

$$\begin{pmatrix} f(t/\delta) \\ g(t/\delta) \end{pmatrix} = \delta^{-\frac{1}{2}} \begin{pmatrix} a^2 & c^2 \\ b^2 & d^2 \end{pmatrix} \begin{pmatrix} f(t) \\ g(t) \end{pmatrix} + \left(\frac{t}{\delta}\right)^{\frac{1}{2}} \begin{pmatrix} ac \\ bd \end{pmatrix}. \tag{5.7}$$

Add Eq. (5.6) to Eq. (5.7) to obtain

$$\begin{pmatrix} 0 \\ 0 \end{pmatrix} = \delta^{-\frac{1}{2}} \begin{pmatrix} 0 & (b+c)^2 \\ (b+c)^2 & 0 \end{pmatrix} \begin{pmatrix} f(t) \\ g(t) \end{pmatrix} + \left(\frac{t}{\delta}\right)^{\frac{1}{2}} \begin{pmatrix} a(b+c) \\ d(b+c) \end{pmatrix}. \tag{5.8}$$

If $b \neq c$, then $f(t) = t^{\frac{1}{2}} \left(\frac{d}{b+c}\right)$ and $g(t) = t^{\frac{1}{2}} \left(\frac{a}{b+c}\right)$. This implies that $\mathcal{C}$ is classical. So suppose that $\mathcal{C}$ is not classical, i.e., $b = c$, implying $b^2 = ad + 1$. This also means that we are assuming that $q \geq 8$. (When $q = 8$ there is considerable collapsing of coefficients, etc., and the GQ is isomorphic to the FTWKB and to the Payne GQ. For example, necessarily $\delta^3 = \delta + 1$, $f(t) = (1+\delta)t^6 + \delta^2 t^4 + \delta t^2$, $g(t) = \delta t^6 + \delta^2 t^4 + (1+\delta)t^2$ and $g(t) = \delta^{\frac{1}{2}} f(t/\delta)$. The collineation under consideration exists with $a = 0$; $b = c = 1$; $d = \delta^5$. Hence we may assume that $q \geq 16$.) Multiply Eq. (5.6) on the left by $(d\delta^{\frac{1}{2}}, a\delta^{\frac{1}{2}})$ to get

$$a\left(f(t) + \delta^{\frac{1}{2}} g\left(\frac{t}{\delta}\right)\right) = d\left(g(t) + \delta^{\frac{1}{2}} f\left(\frac{t}{\delta}\right)\right) = 0 \quad \text{for all } t \in F. \tag{5.9}$$

We conjecture that $(f(t) + \delta^{\frac{1}{2}} g(\frac{t}{\delta}))$ is not a constant times $(g(t) + \delta^{\frac{1}{2}} f(\frac{t}{\delta}))$ unless $m = \pm 1$ or $m = \pm\frac{q}{2}$ (or q is quite small).

Now suppose that $(f(t) + \delta^{\frac{1}{2}} g(\frac{t}{\delta}))$ is indeed a constant times $(g(t) + \delta^{\frac{1}{2}} f(\frac{t}{\delta}))$ in the Subiaco case. The details are a bit cumbersome, but they lead in a very straightforward way to the equation $0 = 1 + \delta^3 + \delta^5 + \delta^8 = (1+\delta^3)(1+\delta^5)$. Recall that $GF(2)(\delta) = GF(q)$ (see Exercise 3.7.2). If $\delta^3 = 1$, then $\delta \in GF(4)$, so for $q \geq 16$ we have a contradiction. If $\delta^5 = 1$, then $\delta \in GF(16)$. So if $q > 16$ we have the desired contradiction.

So let $q = 16$ and $\delta^5 = 1$. After some straighfoward but cumbersome computations (the details are left to the reader) we arrive at $\delta^2 = 1$, which is clearly not possible since $\delta^5 = 1$. So, also for $q = 16$ and $\delta^5 = 1$ we get a contradiction. This completes the proof.

In Chapter 6 we give a much fuller treatment of the Subiaco oval stabilizers in the various cases.

5.4 The Adelaide Examples: $q = 2^e$ with e Even

Putting $k = 1$ and $m = \frac{q-1}{3}$ in Eqs. (4.7) and (4.8) (and always requiring that e be even) gives us the following q-clan functions:

$$f(t) = \frac{1}{[1]}\left\{\left[\frac{1}{6}\right] + \left[\frac{5}{6}\right]t\right\} + \frac{T\left((\beta^{\frac{1}{2}}t + \bar{\beta}^{\frac{1}{2}})^{\frac{q-1}{3}}\bar{\beta}^{\frac{1}{2}}\right)}{[1](t + (\delta t)^{\frac{1}{2}} + 1)^{\frac{q-4}{3}}} + \left(\frac{t}{\delta}\right)^{\frac{1}{2}},$$

and

$$g(t) = \frac{1}{[1]}\left\{\left[\frac{5}{6}\right] + \left[\frac{1}{6}\right]t\right\} + \frac{T\left((\beta^{\frac{1}{2}}t + \bar{\beta}^{\frac{1}{2}})^{\frac{q-1}{3}}\beta^{\frac{1}{2}}\right)}{[1](t + (\delta t)^{\frac{1}{2}} + 1)^{\frac{q-4}{3}}} + \left(\frac{t}{\delta}\right)^{\frac{1}{2}}.$$

If we put $t = \frac{[j+1]}{[j]}$ and use the square bracket function we get

$$f\left(\frac{[j+1]}{[j]}\right) = \frac{\left[\frac{5(j+1)}{6}\right]\left[\frac{j}{6}\right]}{[j]} + \left(\frac{[j+1]}{[j][1]}\right)^{\frac{1}{2}},$$

and

$$g\left(\frac{[j+1]}{[j]}\right) = \frac{\left[\frac{(j+1)}{6}\right]\left[\frac{5j}{6}\right]}{[j]} + \left(\frac{[j+1]}{[j][1]}\right)^{\frac{1}{2}}.$$

We know that G_0 contains a subgroup isomorphic to $C_{q+1} \rtimes C_{2e}$. In the next chapter we will show that the induced oval stabilizer, which is cyclic of order $2e$, is the complete oval stabilizer. This shows that we have the full group G_0. (In [OP02] there is also a proof (depending on significant earlier work) that if $\hat{G}_0$ contains a Sylow 2-subgroup of $PGL(2, q)$, then the herd is classical or FTWKB.)

Chapter 6

The Subiaco Oval Stabilizers

6.1 Algebraic Plane Curves

Definition 6.1.1. *An* algebraic plane curve of degree n *($n \geq 1$) in $PG(2,q)$ is a set of points $C = V(f) = \{(x,y,z) \in PG(2,q)\colon f(x,y,z) = 0\}$, where f is a homogeneous nonzero polynomial of degree n in the variables x, y, z.*

If f is irreducible over $F = GF(q)$, then C is *irreducible*, and if f is irreducible over the algebraic closure $\widehat{F}$ of $F = GF(q)$, then C is *absolutely irreducible.* It turned out that each Subiaco hyperoval is the pointset of an absolutely irreducible degree 10 algebraic curve in $PG(2,q)$. (This was shown in [OKT96] for $q = 2^e$ with $e \not\equiv 2 \pmod 4$ and in [PPP95] for $e \equiv 2 \pmod 4$. Both cases are treated in [Pa98].)

The automorphism group of $PG(2,q)$ is the group $P\Gamma L(3,q)$ induced by the semi-linear transformations of the underlying vector space, which transformations we call *collineations.*

Definition 6.1.2. *The elements of the normal subgroup $PGL(3,q)$ determined by the linear transformations will be called* homographies.

If $\sigma : x \mapsto x^\sigma$ is an automorphism of F, then σ induces a collineation of $PG(2,q)$ called an *automorphic collineation*, as follows: $\sigma : (x,y,z) \mapsto (x^\sigma, y^\sigma, z^\sigma)$. Let $\mathcal{A}$ denote the group of automorphic collineations of $PG(2,q)$, so that $\mathcal{A} \cong Aut(F)$ and $|\mathcal{A}| = e$, if $q = 2^e$. Also, $P\Gamma L(3,q) = PGL(3,q) \rtimes \mathcal{A}$.

If X is a set of points in $PG(2,q)$, the stabilizer $P\Gamma L(3,q)_X$ of X in $P\Gamma L(3,q)$ is called the *collineation stabilizer of* X, while the stabilizer $PGL(3,q)_X$ is called the *homography stabilizer of* X. A set of points in $PG(2,q)$ which is the image under an element of $P\Gamma L(3,q)$ of a set X of points is said to be (projectively) equivalent to X.

Recall that an element $g \in P\Gamma L(3,q)$ is of the form $g : p \mapsto p^\sigma B$, where $p = (x,y,z) \in PG(2,q)$, $B \in GL(3,q)$, and $\sigma \in \mathcal{A}$. The image C^g of an algebraic curve $C = V(f)$ under g is the curve $C^g = \{(x,y,z)^g : (x,y,z) \in C\}$. If we want to write $C^g = V(h)$ for some homogeneous polynomial $h = f^g$, we write

$$f^g(x,y,z) = h(x,y,z) = (f((x,y,z)^{g^{-1}}))^{\sigma}.$$

Indeed, if $p^g = p^{\sigma}B$, then $p^{g^{-1}} = p^{\frac{1}{\sigma}}B^{-\frac{1}{\sigma}}$. So $f^g(p^g) = [f((p^g)^{g^{-1}})]^{\sigma} = f(p)^{\sigma}$ and $f(p) = 0$ if and only if $f^g(p^g) = 0$. At the same time $f(p^{g^{-1}}) = f(p^{\frac{1}{\sigma}}B^{-\frac{1}{\sigma}})$ gives a polynomial in $x^{\frac{1}{\sigma}}, y^{\frac{1}{\sigma}}, z^{\frac{1}{\sigma}}$, so that $(f(p^{g^{-1}}))^{\sigma} = f^{\sigma}(pB^{-1})$ gives a polynomial in x, y, z. Here, if $f(x,y,z) = \sum a_{ij}x^iy^jz^k$, then $f^{\sigma}(x,y,z) = \sum a_{ij}^{\sigma}x^iy^jz^k$. Also note that f_x, f_y, f_z denote the partial derivatives of f with respect to x, y, z, respectively.

Definition 6.1.3. *Let p be a point of the algebraic plane curve $C = V(f)$ of $PG(2,q)$, and let L be a line containing p. Let $p' \in L \setminus \{p\}$; let $p = (x,y,z)$ and let $p' = (x',y',z')$ (we will also say $p(x,y,z)$ and $p'(x',y',z')$).*

If $f(x+tx', y+ty', z+tz') = F(t)$, then the intersection multiplicity *of L and C at p, denoted $m_p(L,C)$, is the multiplicity of the root $t = 0$ of $F(t)$; see* 2.7 *of* [Hi98].

If $F(t)$ is the zero polynomial, then we say that $m_p(L,C) = \infty$. This intersection multiplicity $m_p(L,C)$ is invariant under the action of $P\Gamma L(3,q)$. Let $\hat{F}$ denote the algebraic closure of $F = GF(q)$, so that $\hat{C} = V(f)$ is an algebraic plane curve in $PG(2,\hat{F})$.

Definition 6.1.4. *The* multiplicity *of p on C, denoted $m_p(C)$, is the minimum of $m_p(L,\hat{C})$ for all lines L through p in $PG(2,\hat{F})$. Then p is a* singular *point of C if $m_p(C) > 1$ and a* simple *point of C if $m_p(C) = 1$.*

The curve C is called *singular* or *nonsingular* according as C does or does not have a singular point. A line L of $PG(2,\hat{F})$ containing p is a *tangent* line to C at p or *touches* C at p if $m_p(L,\hat{C}) > m_p(C)$. The point p is singular if and only if $f_x(p) = f_y(p) = f_z(p) = 0$. If p is simple, then $f_x(p)x + f_y(p)y + f_z(p)z = 0$ is the equation of the unique tangent line of C at p.

Theorem 6.1.5 ([Hi98], 2.7, 2.8). *Let $C = V(f)$ be an algebraic plane curve of degree n in $PG(2,q)$. Further, suppose that $f(x,y,z) = \sum_{i=0}^{n} f^{(i)}(x,y)z^{n-i}$, where $f^{(i)}$ is a (homogeneous) polynomial of degree i in the variables x, y. Then*

(i) *If $f^{(0)} = f^{(1)} = \cdots = f^{(m-1)} = 0$ but $f^{(m)} \neq 0, 1 \leq m \leq n$, then C has a point of multiplicity m at $(0,0,1)$.*

(ii) *With m as in (i), there exists $k \leq m$ such that the curve $V(f^{(m)})$ consists of m lines in $PG(2,q^k)$ each of which is a tangent line to C at $(0,0,1)$. Here a line corresponding to a linear factor of $f^{(m)}$ with multiplicity s is counted s times, and is said to have multiplicity s at $(0,0,1)$. These multiplicities are invariant under the action of $P\Gamma L(3,q)$.*

Let $C_1 = V(f_1)$ and $C_2 = V(f_2)$ be algebraic plane curves in $PG(2,q)$ of degrees n_1 and n_2, respectively, and let $p \in C_1 \cap C_2$. Let $\hat{F}$ denote the algebraic closure of $F = GF(q)$, so that $\hat{C}_1 = V(f_1)$ and $\hat{C}_2 = V(f_2)$ are algebraic plane curves of degrees n_1 and n_2 in $PG(2,\hat{F})$. Assume $|\hat{C}_1 \cap \hat{C}_2| < \infty$, that is, $\hat{C}_1$ and $\hat{C}_2$ have no common component. Coordinates are chosen in such a way that $p = (0,0,1)$, $(1,0,0) \notin C_1 \cap C_2$, and such that any line of $PG(2,\hat{F})$ containing $(1,0,0)$ has at most one point in common with $\hat{C}_1 \cap \hat{C}_2$.

Definition 6.1.6. *If f_1 and f_2 are two polynomials in a variable z with coefficients in a field K, of degrees m and n, ($f_1 = a_0 + a_1 z + a_2 z^2 + \cdots + a_m z^m$ and $f_2 = b_0 + b_1 z + b_2 z^2 + \cdots + b_n z^n$), the Resultant $R(f_1, f_2)$ of f_1 and f_2 with respect to z is the determinant of the $(m+n) \times (m+n)$ matrix*

$$\begin{pmatrix} a_0 & a_1 & \dots & a_m & 0 & 0 & \dots & 0 \\ 0 & a_0 & a_1 & \dots & a_m & 0 & \dots & 0 \\ \vdots & & & & & & & \\ 0 & 0 & \dots & a_0 & a_1 & \dots & \dots & a_m \\ b_0 & b_1 & \dots & b_n & 0 & \dots & \dots & 0 \\ 0 & b_0 & b_1 & \dots & b_n & 0 & \dots & 0 \\ \vdots & & & & & & & \\ 0 & \dots & \dots & 0 & b_0 & b_1 & \dots. & b_n \end{pmatrix}$$

of coefficients of these polynomials.

Generalizing slightly, suppose that f_1 and f_2 are polynomials in the variable z over the ring $K[x_1, ..., x_n]$. We can still form the matrix of coefficients with entries a_i and b_j that are polynomials in $x_1, ..., x_n$: the determinant will be a polynomial in $K[x_1, ..., x_n]$, again called the resultant of f_1 and f_2 with respect to z.

Let $R_{x0}(f_1, f_2)$ be the resultant of f_1 and f_2 with respect to x_0; so $R_{x0}(f_1, f_2)$ has degree $n_1 n_2$. If x_1^s is the largest power of x_1 which divides $R_{x0}(f_1, f_2)$, then s is called the *intersection multiplicity* of C_1 and C_2 at p, denoted $I(p, C_1 \cap C_2)$. If $\hat{C}_1$ and $\hat{C}_2$ have a common component, but p does not belong to a common component of $\hat{C}_1$ and $\hat{C}_2$, then delete the common components, which yields curves C_1' and C_2' of $PG(2,q)$, and define $I(p, C_1 \cap C_2) = I(p, C_1' \cap C_2')$. This intersection multiplicity is invariant under the action of $P\Gamma L(3,q)$; see e.g., [Se68]. Also, this definition of intersection multiplicity is consistent with the definition of $m_p(L, C)$.

Theorem 6.1.7 (Theorem of Bézout; see [Se68]). *Let $C_1 = V(f_1)$ and $C_2 = V(f_2)$ be algebraic plane curves of degrees n_1 and n_2, respectively. Let $\hat{F}$ denote the algebraic closure of $F = GF(q)$, so that $\hat{C}_1 = V(f_1)$ and $\hat{C}_2 = V(f_2)$ are algebraic plane curves of degrees n_1 and n_2 in $PG(2,\hat{F})$. If $\hat{C}_1$ and $\hat{C}_2$ have no common component, then*

$$\sum_{p \in \hat{C}_1 \cap \hat{C}_2} I(p, \hat{C}_1 \cap \hat{C}_2) = n_1 n_2. \tag{6.1}$$

Corollary 6.1.8 ([Hi98], 2.8).

$$\sum_{p\in \hat{C}_1\cap\hat{C}_2} m_p(\hat{C}_1)m_p(\hat{C}_2) \le n_1n_2. \tag{6.2}$$

Lemma 6.1.9. *Let C_1 and C_2 be plane algebraic curves each containing p as a simple point, having no common component containing p, each with tangent L at p and such that $m_p(L, C_1) = s$ and $m_p(L, C_2) = t$, where $s \le t$. Then $I(p, C_1 \cap C_2) \ge s$.*

Proof. Coordinates are chosen in such a way that $p = (0, 0, 1)$, $(1, 0, 0) \notin C_1 \cap C_2$, $L = V(x_0)$, and such that any line of $PG(2, \widehat{F})$ containing $(1, 0, 0)$ has at most one point in common with $\hat{C}_1 \cap \hat{C}_2$. Then $C_1 = V(f_1)$ and $C_2 = V(f_2)$, with

$$f_1 = x_0g_1(x_0, x_1, x_2) + x_1^s h_1(x_1, x_2), \tag{6.3}$$
$$f_2 = x_0g_2(x_0, x_1, x_2) + x_1^t h_2(x_1, x_2), \tag{6.4}$$

for some polynomials g_i, h_i, $i = 1, 2$. Then it is clear that x_1^s divides $R_{x_0}(f_1, f_2)$, so that $I(p, C_1 \cap C_2) \ge s$. □

6.2 The Action of $\mathcal{G}_0$ on the $\mathcal{R}_\alpha$

In this section we continue to let ζ be a primitive element for $EGF(q^2) \supseteq F = GF(q)$, $\beta = \zeta^{q-1}$, and $\delta = \beta + \beta^{-1}$.

In Lemma 3.9.2 (with $m = 5$) we observed that

$$\theta = I_1 \circ I_\infty = \theta(id, M \otimes M^{-5}). \tag{6.5}$$

We know (see Eq. (2.30)) that just as $\theta : \mathcal{L}_\gamma \to \mathcal{L}_{\gamma M}$, also $\theta : \mathcal{R}_\alpha \to \mathcal{R}_{\alpha M^{-5}}$. And since $\gamma \mapsto \gamma M$ permutes $\gamma \in PG(1, q)$ in a cycle of length $q + 1$, $\alpha \mapsto \alpha M^{-5}$ permutes the $\alpha \in PG(1, q)$ in a cycle of length $q + 1$ if $q + 1$ is not divisible by 5, and in five cycles of length $(q + 1)/5$ when $q + 1$ is divisible by 5. This proves the following:

Theorem 6.2.1. *Let $\delta = \beta + \beta^{-1}$ where $\beta = \zeta^{q+1}$ for some primitive element ζ for $E = GF(q^2)$. Then*

(i) *If $e \not\equiv 2 \pmod 4$, that is, $q + 1 \not\equiv 0 \pmod 5$, then $I_1 \circ I_\infty$ permutes the $\mathcal{R}_\alpha$ (and hence the ovals $\mathcal{O}_\alpha$) in a cycle of length $q + 1$.*

(ii) *If $e \equiv 2 \pmod 4$ so $q + 1 \equiv 0 \pmod 5$), $I_1 \circ I_\infty$ permutes the $\mathcal{R}_\alpha$ in five cycles of length $(q + 1)/5$.*

Corollary 6.2.2. *If $q = 2^e$ with $e \not\equiv 2 \pmod 4$, there is a unique Subiaco oval. It has a stabilizer of order $2e$ inherited from $\mathcal{G}_0$.*

Proof. The uniqueness follows from part (i) of Theorem 6.2.1, since each two canonical Subiaco GQ are equivalent (see Theorem 4.5.1). Then since $\mathcal{G}_0$ is transitive on all $q+1$ $\mathcal{R}_\alpha$, $|\mathcal{G}_0| = 2e(q^2-1)$, and the kernel $\mathcal{N}$ induces the identity map on the plane of the oval $\mathcal{O}_\alpha$. □

Later we will study more thoroughly this induced group of order $2e$. Now we recall Lemma 3.9.2 in the particular case $m = 2$ and using the 's notation' instead of the 'j notation'

Lemma 6.2.3. *The unique involution of $GQ(\mathcal{C})$ in $\mathcal{G}_0$ fixing $[A(s)]$, $s \in F$, is $I_s = \theta(id, A_s \otimes B_s)$, where*

$$A_s = \begin{pmatrix} 1+\left(\frac{s\delta}{s^2+\delta s+1}\right)^{1/2} & s\left(\frac{\delta}{s^2+\delta s+1}\right)^{1/2} \\ \left(\frac{\delta}{s^2+\delta s+1}\right)^{1/2} & 1+\left(\frac{s\delta}{s^2+\delta s+1}\right)^{1/2} \end{pmatrix}$$

and

$$B_s = \begin{pmatrix} a(s) & b(s) \\ c(s) & a(s) \end{pmatrix}$$

where

$$\begin{aligned} a(s) &= \frac{(s^5+1)(1+\delta+\delta^2)+(s^4+s)(1+\delta)}{(s^2+\delta s+1)^{5/2}}, \\ b(s) &= \frac{\delta^{3/2}s^5+\delta^{1/2}s^4+\delta^{3/2}s+\delta^{1/2}+\delta^{5/2}}{(s^2+\delta s+1)^{5/2}}, \\ c(s) &= \frac{(\delta^{1/2}+\delta^{5/2})s^5+\delta^{3/2}s^4+\delta^{1/2}s+\delta^{3/2}}{(s^2+\delta s+1)^{5/2}}, \end{aligned}$$

$I_s\colon [A(t)] \mapsto [A(\bar{t})]$, *where* $\bar{t} = \dfrac{t(s^2+1)+\delta s^2}{t\delta+s^2+1}$.

It is also useful to recall (for $m = 5$) that

$$I_\infty = \theta\left(id, \begin{pmatrix} 1 & \delta^{1/2} \\ 0 & 1 \end{pmatrix} \otimes B_\infty\right) \text{ where } B_\infty = \begin{pmatrix} 1+\delta+\delta^2 & \delta^{3/2} \\ \delta^{1/2}+\delta^{5/2} & 1+\delta+\delta^2 \end{pmatrix}.$$

Lemma 6.2.4. *$\mathcal{G}_0$ has exactly $q+1$ involutions, each fixing a unique $\mathcal{R}_\alpha$.*

Proof. We know that for each $s \in \tilde{F}$ there is a unique involution I_s fixing $[A(s)]$. Since $q+1$ is odd, any involution would have to fix some line $[A(s)]$. Hence $\mathcal{G}_0$ has just the involutions I_s. For each $s \in \tilde{F}$, $I_s : \mathcal{R}_\alpha \to \mathcal{R}_{\alpha B_s}$ for a special matrix B_s. See Lemma 6.2.3 for B_∞ and for B_s with $s \in F$. In all cases it is easy to see that $B_s \neq I$, $tr(B_s) = 0$, $det(B_s) = 1$. Hence $x^2+1 = (x+1)^2$ is both the minimal and the characteristic polynomial for B_s. This implies that B_s has a 1-dimensional space of eigenvectors belonging to the eigenvalue 1. □

Note that the preceding lemma does not claim that each $\mathcal{R}_\alpha$ is fixed by some involution. When $e \not\equiv 2 \pmod 4$, $\mathcal{G}_0$ is transitive on the entire set of $\mathcal{R}_\alpha{}'s$, so their stabilizers are conjugate in $\mathcal{G}_0$, implying that indeed each $\mathcal{R}_\alpha$ is fixed by a unique involution. However, when $e \equiv 2 \pmod 4$, this will turn out not to be the case.

It is clear that the cases $e \equiv 2 \pmod 4$ and $e \not\equiv 2 \pmod 4$ are different. Hence we adopt a slightly different approach for the case $e \equiv 2 \pmod 4$.

6.3 The Case e ≡ 2 (mod 4)

Let $e \equiv 2 \pmod 4$. Then $\delta = w \in F$ with $w^2 + w + 1 = 0$ satisfies $tr(1/\delta) = w + w^2 = 1$. So we may use $\delta = w$ to define a q-clan (see Observation 4.1.3) but this choice of δ cannot be used in Theorem 6.2.1, for example. Put $L = GF(2)(\delta) = GF(4)$, and $r = [F : L]$, so $e = 2r$ with r odd. Using Lemma 6.2.3, we find that:

$$\begin{aligned}
&\text{(i)} \quad I_\infty = \theta\left(id, \begin{pmatrix} 1 & w^2 \\ 0 & 1 \end{pmatrix} \otimes P\right) : (\alpha, \beta, c) \mapsto \\
&\qquad (\alpha P, (w^2\alpha + \beta)P, c + g(\alpha, w)); \quad (\text{and } A_w = \begin{pmatrix} w & w^2 \\ 0 & w \end{pmatrix}). \\
&\text{(ii)} \quad I_1 = \theta(id, P \otimes P) : (\alpha, \beta, c) \mapsto (\beta P, \alpha P, c + \alpha \circ \beta). \\
&\text{(iii)} \quad I_1 \circ I_\infty = \theta\left(id, \begin{pmatrix} 0 & 1 \\ 1 & w^2 \end{pmatrix} \otimes I\right).
\end{aligned} \tag{6.6}$$

It is clear from Eq. (6.6)(iii) that $I_1 \circ I_\infty$ leaves invariant each $\mathcal{R}_\alpha$. It is also easy to check that $\left(\begin{smallmatrix} 0 & 1 \\ 1 & w^2 \end{smallmatrix}\right)$ has multiplicative order 5, so that $I_1 \circ I_\infty$ is a collineation of order 5. Clearly $Gal(F/L)$ provides a group of order r stabilizing each $\mathcal{R}_\alpha$ with $\alpha \in PG(1, 4)$, and the kernel provides an additional factor of order $q - 1$. This proves the following (use $|\mathcal{G}_0| = 4r(q^2 - 1)$).

Lemma 6.3.1. *$\langle I_1 \circ I_\infty, Gal(F/L), \mathcal{N}\rangle$ is a group of order $5r(q-1)$ stabilizing $\mathcal{R}_\alpha$ for each $\alpha \in PG(1, 4)$. Hence the orbit containing such an $\mathcal{R}_\alpha$ has length at most* $4(q + 1)/5$. □

With $\delta = w$, the details of Theorem 3.10.3 look a great deal simpler. If $\sigma \in Aut(F)$ with $\sigma : x \mapsto x^{2^j}$, we say σ is **even** or **odd** according as j is even or odd. Using Eq. (3.32) we obtain

Lemma 6.3.2. *The solutions $\bar{0}$ to Eq.* (3.32) *are as follows:*

(i) *for σ even, $\bar{0} \in \{0, w\}$;*

(ii) *for σ odd, $\bar{0} \in \{1, w^2\}$.*

Proof. The calculations are straightforward starting with Eq. (3.33) and using induction. □

The group $\mathcal{H}/\mathcal{N}$ can be given explicitly.

Theorem 6.3.3. *Each special (semi-linear) element of $\mathcal{H}$ is one of the following:*

(i) *for even σ and $\bar{0} = 0$, $\theta = \theta(\sigma, I \otimes I)$;*

(ii) *for even σ and $\bar{0} = w$, $\theta = \theta\left(\sigma, \begin{pmatrix} 1 & w^2 \\ 0 & 1 \end{pmatrix} \otimes P\right)$;*

(iii) *for odd σ and $\bar{0} = 1$, $\theta = \theta\left(\sigma, \begin{pmatrix} 1 & 1 \\ 0 & w \end{pmatrix} \otimes \begin{pmatrix} w & w^2 \\ w^2 & w \end{pmatrix}\right)$;*

(iv) *for odd σ and $\bar{0} = w^2$, $\theta = \theta\left(\sigma, \begin{pmatrix} 1 & w \\ 0 & w \end{pmatrix} \otimes \begin{pmatrix} w^2 & w \\ w & w^2 \end{pmatrix}\right)$.*

Corollary 6.3.4. *Each element of $\mathcal{H}$ leaves invariant $\mathcal{R}_{(1,1)}$.*

Proof. For each of the $\theta(\sigma, A \otimes B) \in \mathcal{H}$ given above, each column sum of B is 1. So $(1,1)B = (1,1)$. □

Theorem 6.3.5. *The stabilizer in $\mathcal{G}_0$ of $\mathcal{R}_{(1,1)}$ is $\cup\{\mathcal{H}\cdot I_s : s \in \tilde{L}\}$ of order $(q-1)10e$, so the $\mathcal{G}_0$-orbit Ω_1 containing $\mathcal{R}_{(1,1)}$ has length $(q+1)/5$.*

Proof. Since $\mathcal{G}_0 = \cup\{\mathcal{H} \cdot I_s : s \in \tilde{F}\}$ (see Lemma 6.2.3: $I_s : [A(\infty)] \mapsto [A(\frac{s^2+1}{w})]$, and $\mathcal{H}$ leaves $\mathcal{R}_{(1,1)}$ invariant, we need only determine which I_s leave $\mathcal{R}_{(1,1)}$ invariant. By Lemma 6.2.3, for $\sigma \in F$, $I_s = \theta(id, A_s \otimes B_s)$ where B_s is given in Lemma 6.2.3. Then I_s leaves $\mathcal{R}_{(1,1)}$ invariant iff B_s has column sums equal, i.e., iff $b(s) = c(s)$. Using Lemma 6.2.3 it is easy to see that this holds iff $ws^4 + ws = 0$ iff $s \in GF(4)$. As $\mathcal{H} = \mathcal{H}I_\infty$, the result follows. □

Theorem 6.3.6. *If $e \equiv 2 \pmod 4$, $\mathcal{G}_0$ has two orbits on the $\mathcal{R}_\alpha$, the short orbit Ω_1 of length $(q+1)/5$ that contains $\mathcal{R}_{(1,1)}$ and a long orbit Ω_2 of length $4(q+1)/5$.*

Proof. Since $\mathcal{G}_0$ acts on Ω_1 and $|\Omega_1| = (q+1)/5$ is odd, each of the $q+1$ involutions $I_s \in \mathcal{G}_0$ must fix some $\mathcal{R}_\alpha \in \Omega_1$. Hence by Lemma 6.2.4 no I_s can fix any $\mathcal{R}_\beta$ not in Ω_1. Suppose Ω_2 is any $\mathcal{G}_0$-orbit on the $\mathcal{R}_\alpha$'s other than Ω_1. The stabilizer of any $\mathcal{R}_\beta \in \Omega_2$ must have odd order, implying that 4, the highest power of 2 dividing $|\mathcal{G}_0|$, must divide $|\Omega_2|$. By Theorem 6.2.1, we see that $|\Omega_2| = k(q+1)/5$ with $1 \le k \le 4$. Since $(q+1)/5$ is odd, it must be that $k = 4$. □

At this point we do not have an $\alpha \in PG(1,q)$ for which $\mathcal{R}_\alpha \in \Omega_2$. And with $\delta = w$, the next theorem seems to be the best we can do. First, however, we need a general lemma.

Lemma 6.3.7. *The general (monic) irreducible quadratic polynomial over $GF(4)$ has the form: $d(x) = x^2 + px + p^2q$, where $p = 1, w$ or w^2 and $q = w$ or w^2. Also $c(s) = s^5+s^4+w^2s+1 = d(s)[s^3+(1+p)s^2+(p+p^2q^2)s+1+p^2q^2]+ws+1+p+p^2q$. This shows that $c(s)$ is irreducible over $GF(4)$. Hence $c(s) = 0$ for some $s \in F$ iff $5|e$.*

Theorem 6.3.8. *With $e \equiv 2 \pmod 4$ and $\delta = w$ with $w^2 + w + 10$, $\mathcal{R}_{(0,1)} \in \Omega_2$ iff $5 \not| e$.*

Proof. We know that $\mathcal{R}_{(0,1)} \in \Omega_1$ iff $\mathcal{R}_{(0,1)}$ is fixed by some involution I_s, $s \in \tilde{F}$. Clearly I_∞ does not leave $\mathcal{R}_{(0,1)}$ invariant (cf. Lemma 6.2.3). For $s \in F$, I_s leaves $\mathcal{R}_{(0,1)}$ invariant iff $c(s) = 0$, where $c(s) = s^5 + s^4 + w^2 s + 1$. □

The stabilizer of $\mathcal{O}_{(1,1)}$

We will use the normalization as in Observation 1.6.6. Use Section 2.8 to compute the collineations stabilizing the oval

$$\mathcal{O}_\alpha = \{(\gamma_t, g(\alpha, t)) \in PG(2,q) : t \in \tilde{F}\}$$

that are induced by elements of $\mathcal{G}_0$ stabilizing $\mathcal{R}_\alpha$.

6.3.9. Let $q = 2^e$, $e \equiv 2 \pmod 4$; $\delta = \omega$, where $w^2 + w + 1 = 0$. Put $\alpha = (1,1)$. Then

$$\mathcal{O}_{(1,1)} = \{(1, t, h(t)) \in PG(2,q) : t \in F\} \cup \{(0,1,0)\},$$

where $h(t) = \frac{w^2(t^4+t)}{t^4+w^2t^2+1} + t^{\frac{1}{2}}$.

The stabilizer of $\mathcal{O}_{(1,1)}$ induced by $\mathcal{G}_0$ has order $10e$ (since the kernel of the action indicated by Theorem 6.3.5 is $\mathcal{N}$). In Theorem 6.3.3 put $\sigma = 2$ and $\bar{0} = 1$ to get the collineation $\psi = \hat{\theta}\left(2, \begin{pmatrix} 1 & 1 \\ 0 & w \end{pmatrix} \otimes \begin{pmatrix} w & w^2 \\ w^2 & w \end{pmatrix}\right)$ of $PGL(2,q)$ fixing $\mathcal{O}_{(1,1)}$ and defined on $PG(2,q)$ by $\psi : (x,y,z) \mapsto (x^2, x^2 + w^2 y^2, x^2 + w z^2)$.

We may now compute:

$$\psi^2 = \hat{\theta}\left(2^2, \begin{pmatrix} 1 & w^2 \\ 0 & 1 \end{pmatrix} \otimes P\right), \quad \psi^3 = \hat{\theta}\left(2^3, \begin{pmatrix} 1 & w \\ 0 & w \end{pmatrix} \otimes \begin{pmatrix} w^2 & w \\ w & w^2 \end{pmatrix}\right),$$

$$\psi^4 = \hat{\theta}(2^4, I \otimes I), \qquad \psi^e = \hat{\theta}\left(id, \begin{pmatrix} 1 & w^2 \\ 0 & 1 \end{pmatrix} \otimes P\right),$$

$$\psi^{2e} = (id, I \otimes I),$$

$$\theta = \hat{I}_\infty = \hat{\theta}\left(id, \begin{pmatrix} 1 & w^2 \\ 0 & 1 \end{pmatrix} \otimes \begin{pmatrix} 0 & 1 \\ 1 & 0 \end{pmatrix}\right),$$

$$\psi^e : (x,y,z) \mapsto (x, y + wx, z + w^2 x),$$

and

$$\hat{I}_1 = \hat{\theta}(id, P \otimes P) = \phi : (x,y,z) \mapsto (y,x,z).$$

Now $\theta \circ \phi$ has order 5. $\phi \circ \psi \neq \psi \circ \phi$, while $\theta \circ \psi = \psi \circ \theta$.

$$\psi^{-1} = \hat{\theta}\left(2^{-1}, \begin{pmatrix} 1 & w \\ 0 & w \end{pmatrix} \otimes \begin{pmatrix} w^2 & w \\ w & w^2 \end{pmatrix}\right).$$

It is straightforward to check that

$$\psi^{-1} \circ (I_1 \circ I_\infty) \circ \psi = (I_1 \circ I_\infty)^3.$$

So

$$\langle \phi, \theta, \psi \rangle \geq \langle \theta \circ \phi \rangle \rtimes \langle \psi \rangle \cong C_5 \rtimes C_{2e}.$$

Since the stabilizer of $\mathcal{O}_{(1,1)}$ in $\mathcal{G}_0$, modulo the kernel $\mathcal{N}$, has order $10e$, the stabilizer of $\mathcal{O}_{(1,1)}$ induced by $\mathcal{G}_0$ is isomorphic to $C_5 \rtimes C_{2e}$ and has generators $\hat{I}_1$, $\hat{I}_\infty$ and ψ.

Exercise. Show that $h(t+w) = h(t) + w^2$ and use it to show that $\hat{I}_\infty$ does preserve the oval $\mathcal{O}_{(1,1)}$.

Stabilizer of $\mathcal{O}_{(0,1)}$

Let $q = 2^e$, $e \equiv 2 \pmod 4$, $\delta = w$ where $w^2 + w + 1 = 0$, and put $\alpha = (0,1)$. $\mathcal{O}_{(0,1)} = \{(1,t,h(t)) : t \in F\} \cup \{(0,1,0)\}$, where $h(t) = \frac{t^3+t^2+w^2t}{t^4+w^2t^2+1} + wt^{\frac{1}{2}}$. By Theorem 6.3.6, if $5 \not| e$, $\mathcal{R}_{(0,1)} \in \Omega_2$, so $\mathcal{O}_{(0,1)}$ has an induced stabilizer of order $5e/2$ generated by $\hat{I}_1 \circ \hat{I_\infty}$ and $Gal(F/L)$, where $L = GF(4)$.

Exercise. Show $h((t+w)^{-1}) = (h(t)+w)(t+w)^{-1}$. Use this to show that $\hat{I}_\infty \circ \hat{I}_1 : (x,y,z) \mapsto (wx+y, x, z+wx)$ stabilizes $\mathcal{O}_{(0,1)}$.

By Equation (iii) of (6.6),

$$\hat{I}_1 \circ \hat{I}_\infty = \hat{\theta}(id, \left(\begin{smallmatrix} 0 & 1 \\ 1 & w^2 \end{smallmatrix}\right) \otimes I).$$

Here $A = \left(\begin{smallmatrix} a4 & a2 \\ a3 & a1 \end{smallmatrix}\right) = \left(\begin{smallmatrix} 0 & 1 \\ 1 & w^2 \end{smallmatrix}\right)$, so by Section 2.8, $C = 0$, $D = P$, $E = A_w = \left(\begin{smallmatrix} w & w^2 \\ 0 & w \end{smallmatrix}\right)$, $\alpha E \alpha^T = w$. So

$$(1,t,h(t)) \mapsto ((1,t)\left(\begin{smallmatrix} 0 & 1 \\ 1 & w \end{smallmatrix}\right), g((0,1),t) + (1,t)\left(\begin{smallmatrix} 0 \\ w \end{smallmatrix}\right)) = (t, 1+wt, h(t)+wt),$$

i.e., $\hat{\theta}(id, \left(\begin{smallmatrix} 0 & 1 \\ 1 & w^2 \end{smallmatrix}\right) \otimes I) : (x,y,z) \mapsto (y, x+wy, z+wy)$.

Note. $\hat{I}_1 \circ \hat{I}_\infty = (\hat{I}_\infty \circ \hat{I}_1)^{-1}$.

We have essentially completed the proof of the following theorem.

Theorem 6.3.10. *The complete stabilizer of* $\mathcal{O}_{(0,1)}$ *induced by* $\mathcal{G}_0$ *when* $e \equiv 2 \pmod 4$ *and* $5 \not| e$ *is generated by*

(i) $\hat{\theta}_\sigma : (x,y,z) \mapsto (x^\sigma, y^\sigma, z^\sigma)$, $\sigma \in Gal(F/L)$;

(ii) $(\widehat{I_1 \circ I_\infty}) : (x,y,z) \mapsto (y, x+wy, z+wy)$.

Here $\hat{\theta}_\sigma$ *commutes with* $\hat{I}_1 \circ \hat{I}_\infty$, *so the complete induced group is cyclic of order* $5e/2$.

When $5|e$, $\mathcal{R}_{(0,1)}$ belongs to the short orbit and we need to choose a different $\alpha \in \Omega_2$. This is done in the next section.

6.4 The Case e ≡ 10 (mod 20)

In this section $\delta = \beta + \beta^{-1}$ is chosen as at the beginning of Section 6.2, and $q = 2^e$ with $e \equiv 2 \pmod 4$. (After the first few results we will also assume that $e \equiv 10 \pmod{20}$ so that 5 divides both $q+1$ and e.)

Recall $\theta = I_1 \circ I_\infty = \theta(id, M \otimes M^{-5})$, where $M = \begin{pmatrix} 0 & 1 \\ 1 & \sqrt{\delta} \end{pmatrix}$, and more generally (see Section 3.7)

$$M^j = \frac{1}{\delta^{\frac{1}{2}}} \begin{pmatrix} [\frac{j-1}{2}] & [\frac{j}{2}] \\ [\frac{j}{2}] & [\frac{j+1}{2}] \end{pmatrix}. \tag{6.7}$$

Theorem 6.4.1. *When $e \equiv 2 \pmod 4$, $\mathcal{R}_\alpha \in \Omega_2$ for $\alpha = (1, \delta^{\frac{1}{2}})$.*

Proof. Since the orbit of $\mathcal{R}_{(1,1)}$ under $I_1 \circ I_\infty$ is Ω_1, $\mathcal{R}_{(1,\delta^{\frac{1}{2}})} \in \Omega_1$ if and only if there is an integer j for which $(1,1)M^{-5j} \equiv (1, \delta^{\frac{1}{2}})$. This is iff

$$([\tfrac{-5j-1}{2}] + [\tfrac{-5j}{2}], [\tfrac{-5j}{2}] + [\tfrac{-5j+1}{2}]) \equiv (1, \delta^{\frac{1}{2}}) = (1, [\tfrac{1}{2}])$$

iff

$$[\tfrac{-5j}{2}] + [\tfrac{-5j+1}{2}] = [\tfrac{1}{2}]([\tfrac{-5j-1}{2}] + [\tfrac{-5j}{2}])$$

iff

$$[\tfrac{-5j}{2}] + [\tfrac{-5j+1}{2}] = [\tfrac{-5j}{2}] + [\tfrac{-5j-2}{2}] + [\tfrac{-5j+1}{2}] + [\tfrac{-5j-1}{2}]$$

iff

$$[\tfrac{-5j-2}{2}] = [\tfrac{-5j-1}{2}]$$

iff

$$\tfrac{-5j-2}{2} \equiv \pm(\tfrac{-5j-1}{2}) \bmod q+1$$

iff

$$-2 \equiv -1 \pmod{q+1} \text{ or } -5j-2 \equiv 5j+1 (\bmod q+1)$$

iff

$$10j + 3 \equiv 0 \pmod{q+1}.$$

When $e \equiv 2 \pmod 4$, $q+1 \equiv 0 \pmod 5$, so $10j+3 \equiv 0 \pmod{q+1}$ implies that $3 \equiv 0 \pmod 5$. Impossible! So $\mathcal{R}_{(1,\delta^{\frac{1}{2}})} \in \Omega_2$ when $e \equiv 2 \pmod 4$. □

Corollary 6.4.2. *Let $e = 2r$, r odd, and put $\alpha = (1, \delta^{\frac{1}{2}})$. Then*

$$\begin{aligned} \mathcal{O}_\alpha &= \{(1, t, h(t)) : t \in F\} \cup \{(0,1,0)\} \in \Omega_2, \text{ where} \\ h(t) &= \frac{\delta^2 t^4 + \delta^5 t^3 + \delta^2 t^2 + \delta^3 t}{t^4 + \delta^2 t^2 + 1} + (t/\delta)^{\frac{1}{2}}. \end{aligned}$$

At this point we assume that $e \equiv 10 \pmod{20}$ and that $\alpha = (1, \delta^{\frac{1}{2}})$. Since $(\beta^{\frac{q+1}{5}})^5 = 1$ and $\beta^{\frac{q+1}{5}} \neq 1$, $\beta^{\frac{q+1}{5}}$ is a root of $x^4+x^3+x^2+x+1=0$, from which it follows that $w = [\frac{q+1}{5}]$ satisfies $w^2+w+1=0$. Put $W_1 = [\frac{q-4}{5}]$ and $W_2 = [\frac{q+6}{5}]$. Then $W_1 + W_2 = w\delta$ and $W_1 W_2 = w^2 + \delta^2$. So W_1 and W_2 are the two roots in F of $x^2 + w\delta x + w^2 + \delta^2 = 0$.

It is clear that $\phi = (I_1 \circ I_\infty)^{\frac{q+1}{5}} = \theta(id, M^{\frac{q+1}{5}} \otimes I) = \theta(id, A \otimes I)$, where $A^{(2)} = \frac{1}{\delta}\begin{pmatrix} W_1 & w \\ w & W_2 \end{pmatrix}$, induces a linear collineation $\hat{\phi}$ of $PG(2,q)$ stabilizing $\mathcal{O}_\alpha$, and having order 5. Using Theorem 2.8.1 we obtain the following description of $\hat{\phi}$:

$$\hat{\phi} : (\gamma_t, g(\alpha,t)) \mapsto \left(\gamma_t \cdot \frac{1}{\delta}\begin{pmatrix} W_1 & w \\ w & W_2 \end{pmatrix}, g(\alpha,t) + \gamma_t \begin{pmatrix} \alpha C \alpha^T \\ \alpha E \alpha^T \end{pmatrix}\right), \qquad (6.8)$$

where $C = \frac{W_1}{\delta} A_{\frac{w}{W_1}}$, $E = \frac{w}{\delta} A_{\frac{W_2}{w}}$.

In principle we could carry these computations a bit further, but the results do not particularly reward the effort (see SN).

6.5 Subiaco Hyperovals: The Various Cases

For $q \in \{2,4,8\}$, each Subiaco hyperoval is a *regular* hyperoval (i.e., a conic plus its nucleus). (See [Hi98], Sec. 8.4, but keep in mind that the word "oval" in [Hi98] corresponds to the word "hyperoval" in our language.) When $q = 2$, the homography stabilizer has order 24 and is isomorphic to $\mathcal{S}_4$. Since $Aut(F) = \{1\}$ in this case, the collineation stabilizer also has order 24. When $q = 4$, the homography stabilizer has order 360 and is isomorphic to A_6. Hence the collineation stabilizer has order 720 and is isomorphic to $A_6 \rtimes C_2 \cong \mathcal{S}_6$. When $q = 8$, the homography stabilizer has order 504 and is isomorphic to $PGL(2,8)$. Hence the collineation stabilizer has order 1512 and is isomorphic to $P\Gamma L(2,8)$. (See [Hi79], 8.4).

For $q > 8$, no Subiaco hyperoval is regular. However, for the first few values of $q \geq 16$ there is some overlap with previously known constructions. When $q = 16$, a Subiaco hyperoval is a Lunelli–Sce hyperoval ([LS58]). The homography stabilizer has order 36 and is isomorphic to $C_3^2 \rtimes C_4$, while the collineation stabilizer of order 144 is isomorphic to $C_2 \times (C_3^2 \rtimes C_8)$. This follows from results in [PC78], and it is also in [Ko78]. The original source for the collineation stabilizer is [Ha75], but the given group of collineations was not shown to be the entire stabilizer. (More recently this situation has been thoroughly investigated without the use of computers in some work of C. M. O'Keefe and T. Penttila [OKP91].)
If $q = 32$, the unique Subiaco hyperoval is a Payne hyperoval (see Section 8.2). Hence the homography stabilizer has order 2, while the collineation stabilizer of order $2e$ is isomorphic to C_{2e}.

If $q = 64$, then the two projectively distinct Subiaco hyperovals are the two Penttila–Pinneri irregular hyperovals ([PP94]). The homography stabilizer is either

C_5 of order 5 or D_{10} of order 10, and the respective collineation stabilizers have orders 15 and 60 and are isomorphic to C_{15} and $C_5 \rtimes C_{12}$, respectively.

If $q = 128$ or 256, then the projectively unique Subiaco hyperoval was discovered by Penttila and Royle ([PR95]), with homography stabilizer C_2 in each case and collineation stabilizers isomorphic to C_{14} or C_{16}, respectively. (In these cases the order of the collineation stabilizer was obtained with the assistance of a computer.)

In the remainder of this chapter we concentrate on Subiaco ovals and hyperovals with $q > 64$.

First, if $q = 2^e$, with $e \not\equiv 2 \pmod 4$ (so $q + 1 \not\equiv 0 \pmod 5$)), then up to projective equivalence there is only one Subiaco oval. Moreover, the full collineation stabilizer of the corresponding Subiaco hyperoval is the cyclic group of order $2e$ inherited from $\mathcal{G}_0$ (see Section 4.8). For some parts of the proof it is convenient to arrange the situation $e \not\equiv 2 \pmod 4$ into two cases. And the situation with $e \equiv 2 \pmod 4$ is even more involved. Here $q + 1 \equiv 0 \pmod 5$), so there are two types of ovals. With $\delta = w$, the hyperoval $\mathcal{O}^+_{(1,1)}$ is always in the short orbit Ω_1 and has collineation stabilizer $C_5 \rtimes C_{2e}$ of order $10e$. But then $\mathcal{O}^+_{(0,1)}$ is in the long orbit Ω_2 with collineation stabilizer $C_{5e/2}$ if and only if $e \not\equiv 0 \pmod 5$. The last case (i.e., $e \equiv 10 \pmod{20}$) is the most involved, and we have not seen how to use $\delta = w$ to advantage here. So the five cases to receive some individual attention are briefly:

Case 1. $q = 2^e$, e odd. With $\delta = 1$, we may use $\mathcal{O}^+_{(1,1)}$ or $\mathcal{O}^+_{(0,1)}$, as convenient. The collineation stabilizer of the unique hyperoval is C_{2e}.

Case 2. $q = 2^e$, $e \equiv 0 \pmod 4$. There is one hyperoval with collineation stabilizer C_{2e}.

Case 3. $q = 2^e$, $e \equiv 2 \pmod 4$. Put $\delta = w$. Then $\mathcal{O}^+_{(1,1)} \in \Omega_1$ with collineation stabilizer $C_5 \rtimes C_{2e}$.

Case 4. $q = 2^e$, $e \equiv 2 \pmod 4$, $e \not\equiv 0 \pmod 5$. Put $\delta = w$. Then $\mathcal{O}^+_{(0,1)} \in \Omega_2$ with collineation stabilizer $C_{5e/2}$.

Case 5. $q = 2^e$, $e \equiv 10 \pmod{20}$. Choose $\delta = \beta + \beta^{-1}$ as usual. Then $\mathcal{O}^+_{(1,\delta^{1/2})} \in \Omega_2$ (see Theorem 6.4.1) with collineation stabilizer $C_{5e/2}$.

The major goal of this chapter is to show that the complete collineation stabilizer of each Subiaco hyperoval (with $q > 64$) is exactly the group induced by $\mathcal{G}_0$. Along the way we need to show that each Subiaco hyperoval may be represented as an absolutely irreducible algebraic curve of degree 10.

6.6 $\mathcal{O}^+_{(1,1)}$ as an Algebraic Curve

By the previous section we assume that $q = 2^e > 64$. Let $\delta \in F = GF(q)$ satisfy $tr(1/\delta) = 1$. In those cases where $e \not\equiv 2 \pmod 4$ we follow [OKT96], but we do not need to assume that $1+\delta+\delta^2 \neq 0$. Throughout this section $\alpha = (1,1)$ and we note that the corresponding Subiaco hyperoval $\mathcal{O}^+ = \mathcal{O}^+_\alpha$ has the following form:

$$\mathcal{O}^+ = \{(1,t,h(t)) : t \in F\} \cup \{(0,0,1),(0,1,0)\}, \tag{6.9}$$

where

$$h(t) = \frac{\delta^2 t^4 + \delta^2(1+\delta+\delta^2)(t^3+t^2) + \delta^2 t}{(t^2+\delta t+1)^2} + t^{\frac{1}{2}}. \tag{6.10}$$

In this section we find an algebraic plane curve which coincides with $\mathcal{O}^+$ as a set of points in $PG(2,q)$ and begin to investigate its properties. Starting with Eq. (6.10) it is a simple exercise to check that in homogeneous coordinates (x,y,z), the point $(1,t,h(t))$ satisfies

$$\begin{aligned} 0 = \quad & F(x,y,z) = (z^2+xy)(x^2+\delta xy+y^2)^4 \\ & +\delta^4(x^2y^8+x^8y^2) + \delta^4(1+\delta^2+\delta^4)(x^4y^6+x^6y^4). \end{aligned} \tag{6.11}$$

So $F(x,y,z)$ is a homogeneous polynomial of degree 10 and defines an algebraic curve $\mathcal{C}$ in $PG(2,q)$ by

$$\mathcal{C} = V(F(x,y,z)) = \{(x,y,z) \in PG(2,q) : F(x,y,z) = 0\}. \tag{6.12}$$

Lemma 6.6.1. *The curve $\mathcal{C}$ and the hyperoval $\mathcal{O}^+$ coincide as sets of points in $PG(2,q)$.*

Proof. Clearly the curve $\mathcal{C}$ and the hyperoval $\mathcal{O}^+$ coincide on the points (x,y,z) with $x \neq 0$, so we need only check that they coincide on the points $(0,y,z)$. But $F(0,y,z)z^2y^8 = 0$ if and only if $(0,y,z) \in \{(0,0,1),(0,1,0)\}$. □

Let $\widehat{F} = \widehat{GF}(q)$ be the algebraic closure of $GF(q)$ and let $\hat{\mathcal{C}} = V(F(x,y,z))$ denote the corresponding algebraic curve of degree 10 in $PG(2,\widehat{F})$.

Lemma 6.6.2. *The curve $\hat{\mathcal{C}}$ has a unique multiple point $(0,0,1)$ of multiplicity 8, and the two linear factors of $x^2+\delta xy+y^2=0$ (conjugate in a quadratic extension of $GF(q)$) are the equations of the tangents to $\hat{\mathcal{C}}$ at $(0,0,1)$ (each with multiplicity 4).*

Proof. The multiple points of $\hat{\mathcal{C}}$ are determined by the solutions of the following system of equations:

$$\begin{aligned} F(x,y,z) &= 0, \\ F_x(x,y,z) &= y(x^8+\delta^4x^4y^4+y^8) = 0, \\ F_y(x,y,z) &= x(x^8+\delta^4x^4y^4+y^8) = 0, \\ F_z(x,y,z) &= 0. \end{aligned}$$

A multiple point (x,y,z) has $x=0$ if and only if $y=0$, and we have found the multiple point $(0,0,1)$ with multiplicity 8. The factors of $(x^2+\delta xy+y^2)^4=0$ determine the (eight) tangents to $\hat{\mathcal{C}}$ at $(0,0,1)$.

If $x \neq 0$ and $y \neq 0$, then $x^2 + \delta xy + y^2 = 0$. Further, with these conditions holding the following are equivalent:

$$
\begin{aligned}
&F(x, y, z) = 0,\\
&\delta^4(x^2y^8 + x^8y^2) + \delta^4(1 + \delta^2 + \delta^4)(x^4y^6 + x^6y^4) = 0,\\
&\delta^2(xy^4 + x^4y) + \delta^2(1 + \delta + \delta^2)(x^2y^3 + x^3y^2) = 0,\\
&\delta^2xy[y^3 + x^3 + (1 + \delta + \delta^2)(xy^2 + x^2y)] = 0,\\
&\delta^2xy[y(y^2 + \delta xy + x^2) + x(x^2 + \delta xy + y^2) + \delta^2xy(x + y)] = 0,\\
&\delta^4x^2y^2(x + y) = 0,\\
&x = y.
\end{aligned}
$$

Substituting $x = y$ into $x^2 + \delta xy + y^2 = 0$ yields $\delta x^2 = 0$, which is impossible. □

Lemma 6.6.3. *The curve $\mathcal{C}$ is absolutely irreducible.*

Proof. If one of the two tangents to $\mathcal{C}$ at $(0,0,1)$ is a component of $\hat{\mathcal{C}}$, then so is the other, and in this case $x^2 + \delta xy + y^2$ must be a factor of $F(x, y, z)$. Hence $x^2 + \delta xy + y^2$ must divide

$$
\begin{aligned}
&\delta^4(x^2y^8 + x^8y^2) + \delta^4(1 + \delta^2 + \delta^4)(x^4y^6 + x^6y^4)\\
&= \delta^4x^2y^2[y^6 + x^6 + (1 + \delta^2 + \delta^4)(x^2y^4 + x^4y^2)]\\
&= \delta^4x^2y^2[x^2(x^2 + \delta xy + y^2)^2 + y^2(y^2 + \delta xy + x^2)^2 + \delta^4(x^2y^4 + x^4y^2)].
\end{aligned}
$$

This forces $x^2 + \delta xy + y^2$ to divide $\delta^4x^2y^2(y^2 + x^2)$ and hence to divide $x^2 + y^2 = x^2 + \delta xy + y^2 + \delta xy$. But clearly $x^2 + \delta xy + y^2$ does not divide δxy. Thus neither tangent to $\mathcal{C}$ at $(0,0,1)$ is a component of $\hat{\mathcal{C}}$.

As $\hat{\mathcal{C}}$ has a unique singular point, each irreducible factor of $F(x, y, z)$ over $\widehat{F}$ has multiplicity 1. Suppose that the irreducible components of $\hat{\mathcal{C}}$ are $\mathcal{C}_1, \ldots, \mathcal{C}_r$, for some $r > 1$, where $deg(\mathcal{C}_i) = n_i$ and $\mathcal{C}_i$ has multiplicity m_i at $(0,0,1)$. If for some i, we have $m_i = n_i$, then $m_i = n_i = 1$ and the component $\mathcal{C}_i$ is a line, which must therefore be a tangent to $\mathcal{C}_i$, and hence to $\mathcal{C}$, at $(0,0,1)$. This possibility has already been ruled out. Since $n_1 + \cdots + n_r = 10$ and $m_1 + \cdots + m_r = 8$ with $n_i > m_i \geq 0$ for all i, the only possibility is that $r = 2$ and, WLOG, $(n_1, n_2) = (1,9)$, (2,8), (3,7), (4,6) or (5,5), and in each case $m_i = n_i - 1$.

As $(0,0,1)$ is the only singular point of $\hat{\mathcal{C}}$, it is the unique common point of $\mathcal{C}_1$ and $\mathcal{C}_2$. In particular, $(0,0,1)$ is a point on each of $\mathcal{C}_1$ and $\mathcal{C}_2$. Hence $(n_1, n_2) \neq (1,9)$, as otherwise $m_1 = n_1 - 1 = 0$.

First suppose $\mathcal{C}_1$ and $\mathcal{C}_2$ are defined over $GF(q)$. Since any tangent to $\mathcal{C}_i$ at $(0,0,1)$ is a tangent to $\mathcal{C}$ and is therefore not a line of $PG(2,q)$, it follows that any line of $PG(2,q)$ on $(0,0,1)$ meets $\mathcal{C}_i$ in a further point of $PG(2,q)$. Then $|\mathcal{C}_1 \cup \mathcal{C}_2| \geq 2(q+1) + 1 > q + 2|\mathcal{C}|$, a contradiction.

Hence $\mathcal{C}_1$ and $\mathcal{C}_2$ are not defined over $GF(q)$, but rather over some extension $GF(q^s)$, for some $s > 1$. Let σ be a non-identity element of the Galois group $Gal(GF(q^s)/GF(q))$. Then ${\mathcal{C}_1}^\sigma = \mathcal{C}_2$, which implies that $n_1 = n_2 = 5$. By [Hi98], Lemma 2.24, it must be that $|\mathcal{C}_i| \leq 5^2$, so $|\mathcal{C}| \leq 2(25) - 1 = 49$. But $|\mathcal{C}| = q + 2$ and $q > 64$, so the proof is complete. □

Lemma 6.6.4. *If $q > 64$, then $P\Gamma L(3,q)_{\mathcal{O}^+} \leq P\Gamma L(3,q)_{\hat{\mathcal{C}}}$.*

Proof. Let $\theta \in P\Gamma L(3,q)_{\mathcal{O}^+}$. Since ${\mathcal{O}^+}^\theta = \mathcal{O}^+$ and $\mathcal{O}^+ = \mathcal{C}$ as sets of points in $PG(2,q)$, we know that $\mathcal{O}^+ = \mathcal{C}^\theta$. Suppose that $\hat{\mathcal{C}}^\theta \neq \hat{\mathcal{C}}$. Since $\mathcal{O}^+ \subseteq \hat{\mathcal{C}}^\theta \cap \hat{\mathcal{C}}$, and taking multiplicities into account, we see that $\sum_{P\in\hat{\mathcal{C}}^\theta\cap\hat{\mathcal{C}}} m_P(\hat{\mathcal{C}}^\theta) \cdot m_P(\hat{\mathcal{C}}) \geq \sum_{P\in\mathcal{O}^+} m_P(\hat{\mathcal{C}}^\theta) \cdot m_P(\hat{\mathcal{C}}) \geq q + 8 + 8$. By Corollary 6.1.8 and since both $\mathcal{C}$ and $\mathcal{C}^\theta$ are absolutely irreducible, if $\hat{\mathcal{C}}^\theta \neq \hat{\mathcal{C}}$, then $q + 16 \leq 100$, implying that $q \leq 64$. Hence we conclude that $\hat{\mathcal{C}}^\theta = \hat{\mathcal{C}}$ and that $\theta \in P\Gamma L(3,q)_{\hat{\mathcal{C}}}$. □

Lemma 6.6.5. *Let $\theta \in P\Gamma L(3,q)_{\mathcal{O}^+}$. Then θ fixes the point $(0,0,1)$ and fixes the set $\{Q,\hat{Q}\}$ of lines (in a quadratic extension of $PG(2,q)$) determined by the equation $x^2 + \delta xy + y^2 = 0$.*

Proof. First, $\theta \in P\Gamma L(3,q)_{\mathcal{O}^+} \leq P\Gamma L(3,q)_{\hat{\mathcal{C}}}$ by Lemma 6.6.4. Since $(0,0,1)$ is the unique point of multiplicity greater than 1 on $\hat{\mathcal{C}}$, it must be fixed by θ (see part (ii) of Theorem 6.1.5). So the pair $\{Q,\hat{Q}\}$ of tangents to $\mathcal{C}$ at $(0,0,1)$ is also fixed by θ. □

Lemma 6.6.6. *$P\Gamma L(3,q)_{\mathcal{O}^+}$ acts faithfully on $PG(2,q)/(0,0,1) \cong PG(1,q)$ as an element of $P\Gamma L(2,q)$, fixing setwise a pair of (conjugate) points $l,\bar{l}$ in a quadratic extension $PG(1,q^2)$.*

Proof. First, let $PG(2,q)/(0,0,1)$ denote the quotient space of lines on $(0,0,1)$, so that $PG(2,q)/(0,0,1) \cong PG(1,q)$ in the natural way (the line Q corresponds to the point l, etc.). By the preceding lemma, an element $\theta \in P\Gamma L(3,q)_{\mathcal{O}^+}$ acts on $PG(2,q)/(0,0,1) \cong PG(1,q)$ as an element of $P\Gamma L(2,q)$, fixing setwise a pair of (conjugate) points $l,\bar{l}$ in a quadratic extension $PG(1,q^2)$. Further, such an action is faithful since no non-trivial element of $P\Gamma L(3,q)$ is a central collineation with center $(0,0,1)$ (for otherwise, since $(0,0,1) \in \mathcal{O}^+$, such a collineation would be an element of $PGL(3,q)$ fixing $\mathcal{O}^+$, and hence a quadrangle, pointwise). Thus $P\Gamma L(3,q)_{\mathcal{O}^+} \leq P\Gamma L(2,q)_{\{l,\bar{l}\}}$. □

Lemma 6.6.7. $P\Gamma L(2,q)_{\{l,\bar{l}\}} = C_{q+1} \rtimes C_{2e}$.

Proof. Let $h \in PGL(2,q)$ with $|h| = q + 1$. Then $\langle h\rangle$ is a Singer cycle of $PG(1,q)$, so $\langle h\rangle = C_{q+1}$ acts regularly on $PG(1,q)$. Since h is induced by an invertible 2×2 matrix over F with (conjugate) eigenvalues in $GF(q^2) \setminus GF(q)$, we have $\langle h\rangle$ fixing a conjugate pair of points, say $\{p,\bar{p}\}$. As $N_{P\Gamma L(2,q)}(\langle h\rangle)$ permutes the fixed points of $\langle h\rangle$, $N_{P\Gamma L(2,q)}(\langle h\rangle)$ stabilizes the conjugate pair $\{p,\bar{p}\}$. Hence

$N_{P\Gamma L(2,q)}(\langle h\rangle) \leq P\Gamma L(2,q)_{\{p,\bar{p}\}}$. And since $P\Gamma L(2,q)$ acts transitively on conjugate pairs of points, then replacing h by a conjugate if necessary,

$$N_{P\Gamma L(2,q)}(\langle h\rangle) \leq P\Gamma L(2,q)_{\{l,\bar{l}\}},$$

where $\{l,\bar{l}\}$ is the pair $\{Q,\bar{Q}\}$ of conjugate tangent lines (considered as conjugate points in the quotient space) to the hyperoval at $(0,0,1)$ in $PG(2,q)$. By [Ka79] and [Hu67], Theorem II.7.3, we have

$$N_{P\Gamma L(2,q)}(\langle h\rangle) = C_{q+1} \rtimes\ C_{2e}.$$

But the order of $P\Gamma L(2,q)_{\{l,\bar{l}\}}$ is $2(q+1)e$, so by comparing orders we obtain the desired result. □

Note. The proof of Lemma 6.6.7 is general, i.e., it has nothing to do with the special case of Subiaco oval considered in this section. In the Section 6.9 we shall cite this proof.

6.7 The Case e ≡ 0 (mod 4)

Continuing with the material of the preceding section, we specialize to the case $e \equiv 0 \pmod 4$ and show that the known collineation group of order 2e stabilizing the hyperoval $\mathcal{O}^+ = \mathcal{O}^+_{(1,1)}$ is its full collineation stabilizer. We have already seen that $P\Gamma L(3,q)_{\mathcal{O}^+}$ is a subgroup of $C_{q+1} \rtimes\ C_{2e}$ and that it contains C_{2e}. If it contains C_{2e} as a proper subgroup, it must have a non-trivial intersection with the cyclic subgroup C_{q+1}. We will show that $P\Gamma L(3,q)_{\mathcal{O}^+}$contains no non-trivial element of C_{2e}, so that $|P\Gamma L(3,q)_{\mathcal{O}^+}| = 2e$, as claimed.

Aiming for a contradiction, we let $G = C_{q+1} \cap P\Gamma L(3,q)_{\mathcal{O}^+}$ be a non-trivial group.

Lemma 6.7.1. *The group G has a unique fixed line.*

Proof. First we note that $G = C_{q+1} \cap P\Gamma L(3,q)_{\mathcal{O}^+} = C_{q+1} \cap PGL(3,q)_{\mathcal{O}^+}$. Let p be a prime such that p divides $|G|$ and let $g \in G$ have order p. Since $1 \neq g \in PGL(3,q)$, g has at most three fixed lines. Further, since p divides $|G|$ and hence divides $q+1$, so $q \equiv -1 \pmod p$, implying that $q^2+q+1 \equiv 1 \pmod p$. As the number of lines fixed by g is of the form $kp+1 \leq 3$ and $p \geq 3$, clearly g has a unique fixed line. From this it follows that C_{q+1} has a unique fixed line and hence also a unique fixed point. (As q is prime to $q+1$ (whereas $q-1$ is not), so any central collineation in C_{q+1} must be a homology and never an elation, the fixed point and line are not incident.) □

The involution $I_1 = \theta(id, P \otimes P)$ must leave invariant$\mathcal{R}_{(1,1)}$ and induce a collineation of $PG(2,q)$ leaving invariant $\mathcal{O}^+$. Using Theorem 2.8.1 we obtain an homography ρ defined as follows. The homography $\rho : (x,y,z) \mapsto (y,x,z)$ is an elation with center $(1,1,0)$ and axis $[1,1,0]^T$, fixing $\mathcal{O}^+$. Thus $\rho \in P\Gamma L(3,q)_{\mathcal{O}^+} \leq P\Gamma L(2,q)_{\{l,\bar{l}\}}$.

Since $C_{q+1} \rtimes P\Gamma L(2,q)_{\{l,\bar{l}\}}$, so $\rho \in N_{P\Gamma L(2,q)_{\{l,\bar{l}\}}}(C_{q+1})$. It follows that ρ permutes the fixed lines of C_{q+1}, and hence fixes the unique fixed line of C_{q+1}. The fixed lines of ρ are $[0,0,1]^T$ and $[1,1,c]^T$ for $c \in GF(q)$, so the fixed line of C_{q+1} (and hence also the fixed line of G) must be one of these lines.

First suppose that the fixed line of G is $[0,0,1]^T$. Then G fixes $(0,0,1)$ (by Lemma 6.6.5) and also fixes $[0,0,1]^T \cap \mathcal{O}^+ = \{(0,1,0),(1,0,0)\}$. If a generator g of G interchanges $(0,1,0)$ and $(1,0,0)$, then g induces an involution on $[0,0,1]^T$, so g fixes a point on $[0,0,1]^T$, hence g and also G fix a line through $(0,0,1)$, contrary to Lemma 6.6.5. Thus g and hence G fix $(0,1,0)$ and $(1,0,0)$ in addition to $(0,0,1)$, implying that G has more than one fixed line, again contrary to Lemma 6.6.5.

Thus the fixed line of C_{q+1} must be of the form $[1,1,c]^T$. By the proof of Lemma 6.7.1, the fixed line of C_{q+1} must not be incident with the fixed point $(0,0,1)$, implying that $c \neq 0$. Let p be a prime such that p divides $|G|$ and let $g \in G$ have order p. The homography g fixes the pencil $\mathcal{P}$ of conics

$$\mathcal{C}_s : (x+y+cz)^2 + s(x^2+\delta xy+y^2) = 0, \quad s \in \tilde{F}.$$

Since p divides $q+1$, so is odd, and since $\mathcal{C}_0 : (x+y+cz)^2 0$ and $\mathcal{C}_\infty : x^2+\delta xy+y^2 = 0$ are fixed by g, at least one more conic $\mathcal{C}_s$ is fixed by g. Since at least three elements of $\mathcal{P}$ are fixed by g, each element of $\mathcal{P}$ is fixed by g. In particular, $\mathcal{C}_1 : c^2z^2+\delta xy = 0$ is fixed by g. We have

$$\mathcal{C}_1 = \left\{\left(1, t, \left(\frac{\delta^{1/2}}{c}\right) t^{\frac{1}{2}}\right) : t \in F\right\} \cup \{(0,1,0)\}.$$

Note that G also fixes the nucleus $(0,0,1)$ of the conic $\mathcal{C}_1$.

Lemma 6.7.2. *If p is any prime dividing $|G|$, then $p \in \{3,5,7\}$.*

Proof. Let p be a prime dividing $|G|$ and let $g \in G$ have order p. Since $\langle g\rangle \leq C_{q+1}$, $\langle g\rangle$ acts semi-regularly on $PG(2,q) \setminus \{(0,0,1)\}$, as the stabilizer in G of any of these points is trivial. Thus any point in $PG(2,q) \setminus \{(0,0,1)\}$ lies in an orbit of length p. Now $\langle g\rangle$ fixes $\mathcal{C}_1$ and $\mathcal{O}^+$, so fixes $\mathcal{C}_1 \cap \mathcal{O}^+$, which must therefore be a union of orbits of $\langle g\rangle$, each of length p (as $(0,0,1) \notin \mathcal{C}_1$). Hence p divides $|\mathcal{C}_1 \cap \mathcal{O}^+|$.

Next we determine $|\mathcal{C}_1 \cap \mathcal{O}^+|$. Certainly $(0,1,0) \in \mathcal{C}_1 \cap \mathcal{O}^+$. Further,

$$(1,t,\frac{\delta^{1/2}}{c}t^{\frac{1}{2}}) \in \mathcal{O}^+ \leftrightarrow \frac{\delta^{1/2}}{c}t^{\frac{1}{2}} = \frac{\delta^2(t^4+t)+\delta^2(1+\delta+\delta^2)(t^3+t^2}{(t^2+\delta t+1)^2)} + t^{\frac{1}{2}}$$

$$\leftrightarrow \delta^2 t^4 + \delta^2(1+\delta+\delta^2)(t^3+t^2) + \delta^2 t + \left(1+\frac{\delta^{1/2}}{c}\right) t^{\frac{1}{2}}(t^4+\delta^2t^2+1) = 0$$

$$\leftrightarrow \delta^4 t^8 + \delta^4(1+\delta^2+\delta^4)(t^6+t^4) + \delta^4 t^2 + \left(1+\frac{\delta}{c^2}\right) t(t^8+\delta^4t^4+1) = 0.$$

Now this is a polynomial over $GF(q)$ in the variable t of degree at most 9, so has at most nine solutions in $GF(q)$. Thus $|\mathcal{C}_1 \cap \mathcal{O}^+| \leq 10$. Since $q+1$ is odd, and p divides $q+1$, it must be that p is odd and the result follows. □

We are in a good position to complete the proof of the main theorem of this section. If $p = 3$, then 3 divides $q+1$, which happens if and only if $q = 2^e$ where e is odd, contrary to assumption. If $p = 5$, then 5 divides $q+1$, which happens if and only if $e \equiv 2 \pmod 4$, again contrary to assumption. Further, $p = 7$ implies $2^e \equiv -1 \equiv 6 \pmod 7$, but the powers of 2 modulo 7 are $\{1, 2, 4\}$, a contradiction.

Hence we now may conclude that the group $G = C_{q+1} \cap P\Gamma L(3,q)_{\mathcal{O}^+}$ is trivial. By the preceding results, this completes a proof of the following theorem.

Theorem 6.7.3. *Let $q = 2^e$ where $e \equiv 0 \pmod 4$ and $q > 64$. The collineation stabilizer $P\Gamma L(3,q)_{\mathcal{O}^+}$ of the essentially unique Subiaco hyperoval $\mathcal{O}^+$ is a cyclic group of order $2e$. The homography stabilizer of $\mathcal{O}^+$ is a cyclic group of order 2, generated by ρ.*

We would like to give a very explicit description of the collineations leaving invariant $\mathcal{O}^+$. In theory we should be able to do this. However, the computations using $\mathcal{O}_{(1,1)}$ seemed rather unpleasant. On the other hand, while perhaps still too messy to be truly satisfying, the computations with $\mathcal{O}_{(0,1)}$ seem a bit more reasonable. So we start with a description of this oval.

$$\text{(i)} \quad \mathcal{O}_{(0,1)} = \{(1, t, h_{(0,1)}(t)) : t \in F\} \cup \{(0,1,0)\},$$

where (6.13)

$$\text{(ii)} \quad h_{(0,1)}(t) = \frac{(\delta^4+\delta^2)t^3+\delta^3t^2+\delta^2t}{t^4+\delta^2t^2+1} + (t/\delta)^{1/2}.$$

Recall from Section 3.7 that

$$M = \begin{pmatrix} 0 & 1 \\ 1 & \sqrt{\delta} \end{pmatrix}$$

$$M^j = \frac{1}{\delta^{1/2}} \begin{pmatrix} [\frac{j-1}{2}] & [\frac{j}{2}] \\ [\frac{j}{2}] & [\frac{j+1}{2}] \end{pmatrix}.$$

Put $\theta = I_1 \circ I_\infty = \theta(id, M \otimes M^{-5})$. Then one may compute directly that

$$\theta^j = \theta(id, M^j \otimes M^{-5j}) = \theta(id, \frac{1}{\delta} \begin{pmatrix} [\frac{j-1}{2}] & [\frac{j}{2}] \\ [\frac{j}{2}] & [\frac{j+1}{2}] \end{pmatrix} \otimes \begin{pmatrix} [\frac{-5j-1}{2}] & [\frac{-5j}{2}] \\ [\frac{-5j}{2}] & [\frac{-5j+1}{2}] \end{pmatrix}).$$

Observe that

$$(I_1 \circ I_\infty)^{-1/10} = \theta(id, M^{-1/10} \otimes M^{1/2})$$

where $M^{1/2} = \frac{1}{\delta^{1/2}} \begin{pmatrix} [\frac{-1}{4}] & [\frac{1}{4}] \\ [\frac{1}{4}] & [\frac{3}{4}] \end{pmatrix}$. It is then clear that $(I_1 \circ I_\infty)^{-1/10}$ maps $\mathcal{O}_{(1,1)}$ to $\mathcal{O}_{(0,1)}$. Since each involution fixes a unique $\mathcal{O}_\alpha$ and the powers of $I_1 \circ I_\infty$

permute the $q+1$ ovals transitively, each of the ovals $\mathcal{O}_\alpha$ is left invariant by a unique involution. Clearly $I_1 = \theta(id, P \otimes P)$ is the unique involution leaving invariant $\mathcal{O}_{(1,1)}$. It follows that

$$\phi = (I_1 \circ I_\infty)^{1/10} \circ I_1 \circ (I_1 \circ I_\infty)^{-1/10}$$

is the unique involution fixing $\mathcal{O}_{(0,1)}$.

In the following computations put $T = [\frac{1}{5}]$, from which it is easy to check that $T^5 + T^3 + T = \delta$. And since M is symmetric with determinant 1, it follows also that $PMP = M^{-1}$. Then it is amusing to use the tensor product notation to compute as follows:

$$\begin{aligned}
\phi &= \theta(id, M^{\frac{1}{10}} \otimes M^{\frac{-1}{2}}) \circ \theta(id, P \otimes P) \circ \theta(id, M^{\frac{-1}{10}} \otimes M^{\frac{1}{2}}) \\
&= \theta(id, M^{\frac{1}{10}}(PM^{\frac{-1}{10}}P)P \otimes M^{\frac{-1}{2}}(PM^{\frac{1}{2}}P)P) \\
&= \theta(id, M^{\frac{1}{5}}P \otimes M^{-1}P) \\
&= \theta\left(id, \frac{1}{\delta^{1/2}}\begin{pmatrix} [\frac{1}{10}] & [\frac{2}{5}] \\ [\frac{3}{5}] & [\frac{1}{10}] \end{pmatrix} \otimes \begin{pmatrix} 1 & \delta^{1/2} \\ 0 & 1 \end{pmatrix}\right) \\
&= \theta\left(id, \delta^{\frac{-1}{2}}\begin{pmatrix} T^{1/2} & T^2 \\ T^3+T & T^{1/2} \end{pmatrix} \otimes \begin{pmatrix} 1 & \delta^{1/2} \\ 0 & 1 \end{pmatrix}\right).
\end{aligned}$$

Since $\phi = I_s = \theta\left(id, \begin{pmatrix} (1+\bar{0}s)^{1/2} & s\bar{0}^{1/2} \\ \bar{0}^{1/2} & (1+\bar{0}s)^{1/2} \end{pmatrix} \otimes B_s\right)$ for some s, it is easy to obtain $s = [\frac{2}{5}]/[\frac{3}{5}] = T/(T^2+1)$ as a ratio of the off-diagonal entries. Using either Theorem 2.8.1 (or Lemma 6.2.3) it is possible to describe this collineation a little more completely.

In the context of Theorem 2.8.1, $C = \delta^{-1}TA_{T^3}$, and $E = \delta^{-1}(T^6+T^2)A_{(\delta+T^3)^{-1}}$ and $A^{(2)} = \delta^{-1}\begin{pmatrix} T & T^4 \\ T^6+T^2 & T \end{pmatrix}$. So (after multiplying through by δ) we find

$$\hat{\phi}\colon (x,y,z) \mapsto (Tx + (T^6+T^2)y, T^4x + Ty, \delta z + Tz_{T^3}x + (T^6+T^2)z_{(\delta+T^3)^{-1}}y).$$

Here it seems not worth the bother to compute the entries z_{T^3} and $z_{(\delta+T^3)^{-1}}$ of two different matrices in the Subiaco q-clan.

In Theorem 2.8.1 put $\sigma = 2$, $\bar{0} = \delta^{-1}$ to obtain

$$\bar{\theta} = \theta\left(2, \begin{pmatrix} 1 & \delta^{-1/2} \\ 0 & \delta^{-1/2} \end{pmatrix} \otimes \begin{pmatrix} \delta^{-1/4} & \delta^{1/4} \\ \delta^{3/4}+\delta^{-1/4} & \delta^{5/4} \end{pmatrix}\right).$$

With a little more straightforward computation one may put $j = 3/5$ and obtain the following:

$$\psi = \bar{\theta} \circ \theta^{3/5} = \theta\left(2, \frac{1}{\delta}\begin{pmatrix} [\frac{7}{10}] & [\frac{1}{5}] \\ [\frac{3}{10}] & [\frac{4}{5}] \end{pmatrix} \otimes \begin{pmatrix} \delta^{1/4} & \delta^{-1/4} \\ 0 & \delta^{-1/4} \end{pmatrix}\right) \qquad (6.14)$$

So ψ yields a collineation $\hat{\psi}$ of $PG(2,q)$ stabilizing $\mathcal{O}_{(0,1)}$ and having companion automorphism $\sigma = 2$. Here

$$A^{(2)} = \frac{1}{\delta^2}\begin{pmatrix} [\frac{7}{5}] & [\frac{2}{5}] \\ [\frac{3}{5}] & [\frac{8}{5}] \end{pmatrix} \frac{1}{\delta^2}\begin{pmatrix} T^7+T^3+\delta & T^2 \\ T^3+T & T^8 \end{pmatrix} = \begin{pmatrix} a_4^2 & a_2^2 \\ a_3^2 & a_1^2 \end{pmatrix}.$$

Then $C = a_4^2 A_{(a2/a4)}2$, $E = a_3^2 A_{(a1/a3)}2$, and $det(A) = \delta^{-1/2}$, so that

$$\hat{\psi} : (x,y,z) \mapsto ((x^2,y^2)A^{(2)}, \delta^{-1}x^2 + x^2 a_4^2 z_{(a2/a4)}2 + y^2 z_3^2 z_{(a1/a3)}2). \tag{6.15}$$

We know that the complete stabilizer of $\mathcal{O}_{(0,1)}$ is cyclic of order $2e$, and we have a concrete description of a collineation with companion automorphism $\sigma = 2$. Hence for each automorphism σ of F there are two collineations of $PG(2,q)$ stabilizing $\mathcal{O}_{(0,1)}$ and having σ as companion automorphism. If $\hat{\psi}$ is one of them, then $\hat{\psi} \circ \hat{\phi}$ is the other. It follows that $\hat{\psi}$ has order $2e$ and that $(\hat{\psi})^e = \hat{\phi}$.

6.8 The Case e Odd

Suppose that $q = 2^e$ with e odd. This is a special case of $e \not\equiv 2 \pmod 4$, so we continue with our study of the stabilizer of $\mathcal{O} = \mathcal{O}_{(1,1)}$ and the corresponding hyperoval $\mathcal{O}^+$. But in this case we may choose $\delta = 1$, so that

$$\mathcal{O}^+ = \{(1,t,f(t)) : t \in F\} \cup \{(0,1,0),(0,0,1)\}$$

where

$$f(t) = \frac{t^4+t^3+t^2+t}{(t^2+t+1)^2} + t^{\frac{1}{2}}. \tag{6.16}$$

The corresponding algebraic curve $\mathcal{C} = V(F)$ is given by

$$\begin{aligned} F(x,y,z) &= (z^2+xy)(x^2+xy+y^2)^4 + x^2y^8 + x^8y^2 + x^4y^6 + x^6y^4 \\ &= (x^2+y^2+z^2+xy)(x^2+xy+y^2)^4 + x^{10} + y^{10} \\ &= (x^2+xy+y^2)^5 + z^2(x^2+xy+y^2)^4 + x^{10} + y^{10}. \end{aligned} \tag{6.17}$$

As in the case for $e \equiv 0 \pmod 4$, we know that there is a group of order $2e$ stabilizing $\mathcal{O}$ and we want to show that this is the full collineation stabilizer and it is cyclic. First we consider the homography stabilizer.

Lemma 6.8.1. *Let $q = 2^e$ with e odd, and let $\mathcal{O}^+$ be the Subiaco hyperoval described above. Then $PGL(3,q)_{\mathcal{O}^+}$ is a group of order 2 generated by the homography $\rho : (x,y,z) \mapsto (y,x,z)$.*

Proof. Let $\theta \in PGL(3,q)_{\mathcal{O}^+}$, so θ can be written as a 3×3 matrix, which we also denote by θ. By Lemma 6.6.5, θ fixes $(0,0,1)$, so θ^{-1} is of the form

$$\theta^{-1} = \begin{pmatrix} a & b & 0 \\ e & f & 0 \\ g & h & 1 \end{pmatrix}$$

for some $a, b, e, f, g, h \in GF(q)$. Further, (also by Lemma 6.6.5) over the algebraic closure $\overline{F}$ of $GF(q)$, θ fixes $\{(x,y,z) : x^2+xy+y^2=0\}$. Writing out this latter condition yields the following:

$$\begin{aligned}&(ax+by)^2+(ax+by)(ex+fy)+(ex+fy)^2\\&=x^2(a^2+ae+e^2)+xy(af+be)+y^2(b^2+fb+f^2)=\lambda(x^2+xy+y^2)\end{aligned} \tag{6.18}$$

for some $\lambda \in GF(q)$. As the above equation holds for all $x, y \in GF(q)$, it follows that

$$a^2+ae+e^2=af+be=b^2+bf+f^2=\lambda.$$

Since θ fixes $\mathcal{O}^+=\mathcal{C}$, if $F(x,y,z)=0$, then $F((x,y,z)^{\theta^{-1}})=0$ also. Hence we obtain, using Eq. (6.18) for simplification at the first step and substituting for $z^2(x^2+xy+y^2)^4$ using Eq. (6.17) at the second step:

$$\begin{aligned}&F((x,y,z)^{\theta^{-1}})=0\\&\Rightarrow \lambda^5(x^2+xy+y^2)^5+(gx+hy+z)^2\lambda^4(x^2+xy+y^2)^4+(ax+by)^{10}+(ex+fy)^{10}=0\\&\Rightarrow \lambda^5(x^2+xy+y^2)^5+\lambda^4(x^2+xy+y^2)^4(g^2x^2+h^2y^2)\\&\qquad\qquad+\lambda^4((x^2+xy+y^2)^5+x^{10}+y^{10})\\&\quad+(a^5x^5+ab^4xy^4+a^4bx^4y+b^5y^5+e^5x^5+ef^4xy^4+e^4fx^4y+f^5y^5)^2=0\\&\qquad\qquad\qquad\qquad\qquad\qquad\qquad\qquad\forall x,y\in GF(q)\\&\Rightarrow x^{10}(\lambda^5+g^2\lambda^4+a^{10}+e^{10})+x^9y(\lambda^5+\lambda^4)+x^8y^2(\lambda^5+\lambda^4+h^2\lambda^4+a^8b^2+e^8f^2)\\&\qquad+x^6y^4(\lambda^5+g^2\lambda^4+\lambda^4)+x^5y^5(\lambda^5+\lambda^4)+x^4y^6(\lambda^5+h^2\lambda^4+\lambda^4)\\&\quad+x^2y^8(\lambda^5+\lambda^4+g^2\lambda^4+a^2b^8+e^2f^8)+xy^9(\lambda^5+\lambda^4)+y^{10}(\lambda^5+h^2\lambda^4+b^{10}+f^{10})=0\end{aligned}$$

for all $x, y \in GF(q)$, not both zero. Thus each coefficient in the last equation must be zero. In particular, the coefficient of x^9y is $\lambda^5+\lambda^4=\lambda^4(\lambda+1)=0$, implying that $\lambda=0$ or $\lambda=1$. But if $\lambda=0$, then the matrix θ is singular (since the determinant of θ^{-1} is $af+be=\lambda$), which is not possible. Thus $\lambda=1$, and the coefficients of x^6y^4 and x^4y^6 imply that $g=h=0$, respectively. We are left with the following four equations, corresponding to the coefficients x^{10}, x^8y^2, x^2y^8, y^{10}:

$$a^5+e^5=1, \tag{6.19}$$

$$a^4b+e^4f=0, \tag{6.20}$$

$$ab^4+ef^4=0, \tag{6.21}$$

$$b^5+f^5=1. \tag{6.22}$$

Multiplying Eq 6.21 by b we obtain $ab^5+ebf^4=0$, hence (by Eq 6.22) $a(1+f^5)+ebf^4=0$, implying $f^4(af+be)+a=0$ and thus $a=f^4$ (since $det\theta^{-1}=af+be=\lambda=1$). Similarly, multiplying Eq. (6.20) by a yields $b=e^4$.

Substituting for a in Eq. (6.21) gives $f^4(b^4 + e) = 0$, so either $f = 0$, or $e = b^4$. If $f = 0$, then $a = f^4 = 0$ and Eqs. (6.19) and (6.22) yield $b^5 = e^5 = 1$. But $b = e = 1$, since the greatest common divisor $(2^e - 1, 5) = 1$ (as e is odd and an odd power of 2 modulo 5 is never 1). In this case θ is the collineation ρ and it is straightforward to verify that ρ fixes $\mathcal{O}^+$ (see Section 6.7). In a similar way, using Eq. (6.20), it follows that either $e = 0$ or $f = a^4$. If $e = 0$, then analogous arguments show that θ is the identity collineation.

We are left with the case in which $e = b^4$ and $f = a^4$. Since $b = e^4$ and $a = f^4$, it follows that $a^{15} = e^{15} = 1$. But this implies that $a = e = 1$, since the greatest common divisor $(15, q-1)(2^4 - 1, 2^e - 1) = 2^{(4,e)} - 1 = 1$, as e is odd. It follows that $b = f = 1$ and

$$\theta^{-1} = \begin{pmatrix} 1 & 1 & 0 \\ 1 & 1 & 0 \\ 0 & 0 & 1 \end{pmatrix}$$

which is impossible since then the determinant of θ^{-1} would be 0. □

Theorem 6.8.2. *In $PG(2,q)$, where $q = 2^e$, e odd and $e \geq 7$, let $\mathcal{O}^+ = \mathcal{O}^+_{(1,1)}$. Then $P\Gamma L(3,q)_{\mathcal{O}^+}$ is a cyclic group of order $2e$, generated by ρ and the automorphic collineation $(x, y, z) \mapsto (x^2, y^2, z^2)$. The homography stabilizer of $\mathcal{O}^+$ is a cyclic group of order 2, generated by ρ.*

Proof. By Lemma 6.8.1, the homography stabilizer of $\mathcal{O}^+$ is $\langle\rho\rangle$, a cyclic group of order 2. Further, since $f(t)$ has coefficients in $GF(2)$, it follows that $\mathcal{O}^+$ is fixed by the group A of automorphic collineations, so the homography stabilizer of $\mathcal{O}^+$ has index e in the collineation stabilizer. Thus the collineation stabilizer of $\mathcal{O}^+$ is $\langle\rho, A\rangle = \langle\rho\rangle \times A$. This is a cyclic group of order $2e$. □

Corollary 6.8.3. *The collineation stabilizer of a Subiaco hyperoval in $PG(2,q)$, where $q = 2^e$, $q \geq 32$ and e is odd, is a cyclic group of order $2e$. Further, its homography stabilizer is a cyclic group of order 2.*

Proof. The case $q = 32$ was discussed in Section 6.5. Suppose $q \geq 128$. A Subiaco hyperoval is equivalent to the hyperoval $\mathcal{O}^+$ described above, so its collineation (respectively, homography) stabilizer is conjugate in $P\Gamma L(3,q)$ (respectively, $PGL(3,q)$) to the stabilizer of $\mathcal{O}^+$, and the result follows. □

6.9 The case e ≡ 2 (mod 4)

In this section we assume that $e \equiv 2 \pmod 4$, so there are two $\mathcal{G}_0$-orbits on the ovals $\mathcal{O}_\alpha$, $\alpha \in PG(1,q)$. The short orbit Ω_1 has length $(q+1)/5$, and the long orbit Ω_2 has length $4(q+1)/5$. For convenience here we also assume that $\delta = w$, where $w^2 + w + 1 = 0$. In this case we know by 6.3.9 that the oval $\mathcal{O}_{(1,1)}$ belongs to the short orbit Ω_1 and has a collineation stabilizer of order $10e$. From the proof of

Theorem 6.3.6 we know that $\mathcal{O}_\alpha$, (that is, $\mathcal{R}_\alpha$) is in the long orbit Ω_2 precisely if it is fixed by no involution I_s. And we may use the description of B_s in Lemma 6.2.3 to recognize members of Ω_2.

We want to reindex the $\alpha \in PG(1,q)$ by $c \in \tilde{F}$ in the following way. For $c \in \tilde{F}$, put

$$\alpha_c = \left(\frac{w^2 + wc}{1 + \sqrt{c} + wc}, \frac{wc}{1 + \sqrt{c} + wc} \right). \tag{6.23}$$

Clearly $1 + \sqrt{c} + wc \neq 0$ for all $c \in F$. And $\alpha_c = (1,1)$ if and only if $c = \infty$, so $\alpha_\infty \in \Omega_1$. For $c, d \in F$, $\alpha_c \equiv \alpha_d$ if and only if $c = d$. For $c = w$, $\alpha_c = \alpha_w = (0, w^2) \equiv (0,1)$. For $c \in \tilde{F} \setminus \{\infty, w\}$, $\alpha_c \equiv (1, c/(c+w))$. Index the ovals $\mathcal{O}_\alpha$ for $\alpha \in PG(1,q)$ with elements of $\tilde{F}$ so that $\mathcal{O}_c = \mathcal{O}_{\alpha_c}$, $c \in \tilde{F}$. Hence we have the following:

$$\begin{aligned} &\mathcal{O}_\infty = \{(1, t, f_\infty(t)) : t \in F\} \cup \{(0,1,0)\}, where \\ &f_\infty(t) = \frac{w^2 t^4 + w^2 t}{t^4 + w^2 t^2 + 1} + t^{\frac{1}{2}}. \end{aligned} \tag{6.24}$$

Then for $c \in F$ we obtain

$$\begin{aligned} &\mathcal{O}_c = \{(1, t, f_c(t)) : t \in F\} \cup \{(0,1,0)\}, \text{ where} \\ &f_c(t) = \left(\tfrac{w^2+wc}{1+\sqrt{c}+wc}\right)^2 \left(\tfrac{w^2t^4+t^3+t^2}{t^4+w^2t^2+1} + wt^{\frac{1}{2}}\right) \\ &\quad + \left(\tfrac{(w^2+wc)wc}{1+c+w^2c^2}\right) t^{\frac{1}{2}} + \tfrac{w^2c^2}{1+c+w^2c^2} \left(\tfrac{t^3+t^2+w^2t}{t^4+w^2t^2+1} + wt^{\frac{1}{2}}\right) \\ &= \tfrac{1}{1+c+w^2c^2} \left(\tfrac{(1+wc^2)t^4+wt^3+wt^2+wc^2t}{t^4+w^2t^2+1} + (w^2+c+w^2c^2)t^{\frac{1}{2}}\right). \end{aligned} \tag{6.25}$$

This choice of α_c may seem unnecessarily complicated, but at least it has the following interesting consequence.

Lemma 6.9.1. *Let $g \in PGL(3,q)$ be defined by $g : (x,y,z) \mapsto (y, x + wy, z + w^2 y)$. Then g is a collineation of order 5 stabilizing $\mathcal{O}_c$ for every $c \in \tilde{F}$.*

Proof. In Section 6.3 it was observed that $I_1 \circ I_\infty = \theta(id, \left(\begin{smallmatrix} 0 & 1 \\ 1 & w^2 \end{smallmatrix}\right) \otimes I)$ is a collineation of order 5 leaving invariant each $\mathcal{R}_\alpha$. We want to use Theorem 2.8.1 to compute the induced collineation $\hat{\theta}(id, \left(\begin{smallmatrix} 0 & 1 \\ 1 & w^2 \end{smallmatrix}\right) \otimes I)$ of $PG(2,q)$ leaving invariant $\mathcal{O}_{\alpha_c}$ for each $c \in \tilde{F}$. In the notation of Section 2.8 we have the following: $A = \left(\begin{smallmatrix} 0 & 1 \\ 1 & w^2 \end{smallmatrix}\right) = \left(\begin{smallmatrix} a_4 & a_2 \\ a_3 & a_1 \end{smallmatrix}\right)$; $C = 0$, $D = P$, $E = A_w = \left(\begin{smallmatrix} w & w^2 \\ 0 & w \end{smallmatrix}\right)$, $B = I$, $\lambda = \mu = 1$, $\sigma = id$. So $A^{(2)} = \left(\begin{smallmatrix} 0 & 1 \\ 1 & w \end{smallmatrix}\right)$, and

$$\begin{aligned} &\hat{\theta} = \hat{\theta}\left(id, \begin{pmatrix} 0 & 1 \\ 1 & w^2 \end{pmatrix} \otimes I\right) : (\gamma_t, g(\alpha, t)) \mapsto \\ &\left(\gamma_t \begin{pmatrix} 0 & 1 \\ 1 & w \end{pmatrix}, g(\alpha, t) + \gamma_t \begin{pmatrix} 0 \\ \alpha E \alpha^T \end{pmatrix}\right). \end{aligned} \tag{6.26}$$

For $\alpha = \alpha_\infty = (1,1)$, we have, for $t \in F$, $\gamma_t \left(\begin{smallmatrix} 0 & 1 \\ 1 & w \end{smallmatrix}\right) = (t, 1+wt)$ and $\alpha E \alpha^T = (1,1)\left(\begin{smallmatrix} w & w^2 \\ 0 & w \end{smallmatrix}\right)\left(\begin{smallmatrix} 1 \\ 1 \end{smallmatrix}\right) = w^2$. Similarly, for $c \in F$, $\alpha_c E \alpha_c^T = \left(\frac{w+w^2c^2}{1+c+w^2c^2}\right) w + \left(\frac{c+w^2c^2}{1+c+w^2c^2}\right) w^2 + \left(\frac{w^2c^2}{1+c+w^2c^2}\right) w = w^2$. So Eq. (6.26) becomes $\hat{\theta} : (1, t, g(\alpha, t)) \mapsto (t, 1+wt, g(\alpha,t)+tw^2)$. It is now easy to see that $\hat{\theta} : (x,y,z) \mapsto (y, x+wy, z+w^2y)$ for all $(x,y,z) \in PG(2,q)$. Hence for every $c \in \tilde{F}$, $\hat{I}_1 \circ \hat{I}_\infty = \hat{\theta}$ is the map g of the lemma. □

Lemma 6.9.2. *The identification of ovals in* Ω_2.

(i) $\mathcal{O}_\infty$ *is always in the short orbit* Ω_1.

(ii) *For* $c \in F$, $\mathcal{O}_c \in \Omega_2$ *if and only if*

$$p_c(x) = x^5 + (w^2c^2 + w^2)x^4 + (w^2c^2 + 1)x + 1 = 0$$

has no root in F.

(iii) *There is a* $c \in F \setminus GF(4)$ *for which both* $p_c(x) = 0$ *and*

$$p_{(w^2c^2+w^2)}(x) = x^5 + (c^4 + w)x^4 + c^4x + 1 = 0$$

have no root in F.

Proof. We have already seen that (i) holds. For $c = w$, $p_w(x) = x^5 + x^4 + w^2x + 1$. This polynomial appeared earlier in Lemma 6.3.7 and Theorem 6.3.8, where it was shown that it is irreducible over F if and only if $\mathcal{O}_{(0,1)} \in \Omega_2$ if and only if 5 does not divide e. For $w \neq c \in F$, $\alpha_c \equiv (1, c/(c+w))$. So put $d = c/(c+w)$. Then

$$\begin{array}{ll}
\mathcal{O}_c \in \Omega_2 & \\
\Longleftrightarrow & \alpha_c \in \Omega_2 \\
\Longleftrightarrow & (1,d) \in \Omega_2 \\
\Longleftrightarrow & (1,d) \text{ is fixed by no involution } I_s \\
\Longleftrightarrow & (1,d)B_s \not\equiv (1,d) \text{ for all } s \in F, \\
 & \text{where } B_s = \begin{pmatrix} a(s) & b(s) \\ c(s) & a(s) \end{pmatrix} \text{ from Lemma 6.2.3,} \\
\Longleftrightarrow & b(s) + da(s) \neq d(a(s) + dc(s)) \text{ for all } s \in F \\
\Longleftrightarrow & b(s) \neq d^2c(s) \text{ for all } s \in F \\
\Longleftrightarrow & d^2 \neq \frac{b(s)}{c(s)} \frac{s^5+w^2s^4+s+1}{s^5+s^4+w^2s+1} \text{ for all } s \in F \\
\Longleftrightarrow & s^5(1+d^2) + s^4(d^2+w^2) + s(1+d^2w^2) + d^2 + 1 = 0 \\
 & \text{has no solution } s \in F \\
\Longleftrightarrow & s^5 + \left(\frac{d^2+w^2}{1+d^2}\right)s^4 + \left(\frac{1+d^2w^2}{1+d^2}\right)s + 1 = 0 \\
 & \text{has no solution } s \in F \\
\Longleftrightarrow & p_c(s) \neq 0 \text{ for all } s \in F, \text{ where } c = \frac{dw}{1+d}, \text{ that is, } d\frac{c}{c+w}.
\end{array}$$

This completes the proof of (ii).

Hence $p_c(x) = 0$ has no root in F for $4(q+1)/5$ values of $c \in F$. Moreover, $c \mapsto w^2c^2+w^2$ is a bijection on F, so $p_{(w^2c^2+w^2)}(x) = 0$ has no root in F also for at least $4(q+1)/5$ values of c. This implies that both $p_c(x) = 0$ and $p_{(w^2c^2+w^2)}(x) = 0$ have no root for at least $3(q+1)/5$ values of c, from which (iii) follows. □

Define the conic $\mathcal{O} : xy = z^2$, that is

$$\mathcal{O} = \{(1, t, t^{\frac{1}{2}}) : t \in F\} \cup \{(0, 1, 0)\}.$$

Lemma 6.9.3. *There are ovals in each orbit that meet $\mathcal{O}$ in its five points defined over $GF(4)$.*

(i) $\mathcal{O}_\infty \cap \mathcal{O} = PG(2, 4) \cap \mathcal{O}$.

(ii) *For c satisfying condition* (iii) *of Lemma* 6.9.2, $\mathcal{O}_c \in \Omega_2$ *and* $\mathcal{O}_c \cap \mathcal{O} = PG(2, 4) \cap \mathcal{O}$.

Proof. To prove (i) note that $(\mathcal{O}_\infty \setminus \{(0, 1, 0)\}) \cap \mathcal{O} = \{(1, t, f_\infty(t)) : f_\infty(t) = t^{\frac{1}{2}}\}$. Clearly $f_\infty(t) = t^{\frac{1}{2}}$ if and only if $t \in GF(4)$. Similarly, if c satisfies condition (iii) of Lemma 6.9.2, then $\mathcal{O}_c \in \Omega_2$ and $(\mathcal{O}_c \setminus \{(0, 1, 0)\}) \cap \mathcal{O}\{(1, t, f_c(t)) : f_c(t) = t^{\frac{1}{2}}\}$. Using Eq. (6.25) we may compute that $f_c(t) = t^{\frac{1}{2}}$ if and only if

$$w^2t(t^3 + 1)(t^5 + (c^4 + w)t^4 + c^4t + 1) = 0,$$

proving (ii). □

For each $c \in \tilde{F}$ we may define the hyperoval $\mathcal{O}_c^+ = \mathcal{O}_c \cup \{(0, 0, 1)\}$, and with a natural abuse of language refer to $\mathcal{O}_c^+$ belonging to Ω_1 or Ω_2. For example, $\mathcal{O}_\infty^+ \in \Omega_1$. And for $c = c_0$ satisfying condition (iii) of Lemma 6.9.2, $\mathcal{O}_{c_0}^+ \in \Omega_2$. For the remainder of this section we fix $c = c_0$ to be such an element.

Lemma 6.6.3 holds for all e, so $\mathcal{O}_\infty^+$ is already known to be the pointset of an absolutely irreducible degree 10 algebraic curve in $PG(2, q)$. For $c = c_0$ we now establish the same result for $\mathcal{O}_c^+$. However, as the same proof works for both cases, without any special effort we obtain another proof for $\mathcal{O}_\infty^+$ when $e \equiv 2 \pmod 4$.

Lemma 6.9.4. *For both $c = \infty$ and $c = c_0$, $\mathcal{O}_c^+$ is the pointset of an absolutely irreducible degree* 10 *algebraic curve $\mathcal{C}$ in $PG(2, q)$ having* $(0, 0, 1)$ *as its only singular point. The multiplicity of* $(0, 0, 1)$ *for $\mathcal{C}$ is* 8, *and in $PG(2, q^2)$ there are two tangent lines for $\mathcal{C}$ at* $(0, 0, 1)$. *The tangent lines are $Q = [\eta, 1, 0]^T : \eta x + y = 0$ and $\bar{Q} = [\eta^q, 1, 0]^T : \eta^q x + y = 0$, where $\eta \in GF(q^2) \setminus GF(q)$ satisfies $\eta^{q+1} = 1$ and $\eta^q + \eta = w$.*

Proof. For $c \in \{\infty, c_0\}$, write $d = c/(c + w)$, so $d \in \{1, c_0/(c_0 + w)\}$. In particular, we are avoiding the case $c = w$ with $d = \infty$. Put $m_c = \left(\frac{w^2+wc}{1+\sqrt{c}+wc}\right)^2$. So $m_c \neq 0$, $m_\infty = 1$, and $\alpha_c = (m_c)^{1/2}(1, d)$. Then $\mathcal{O}_c^+ = \{(1, t, f_c(t)) : t \in F\} \cup \{(0, 1, 0), (0, 0, 1)\}$ is projectively equivalent to $\overline{\mathcal{O}}_c^+ = \{(1, t, f_c(t)/m_c) : t \in F\} \cup \{(0, 1, 0), (0, 0, 1)\}$.

Here

$$\frac{f_c(t)}{m_c} = f(t,d) = \left(\frac{w^2t^4+t^3+t^2}{t^4+w^2t^2+1} + wt^{\frac{1}{2}}\right) + dt^{\frac{1}{2}} + d^2\left(\frac{t^3+t^2+w^2t}{t^4+w^2t^2+1} + wt^{\frac{1}{2}}\right)$$
$$= \frac{w^2t^4+(1+d^2)(t^3+t^2)+w^2d^2t}{t^4+w^2t^2+1} + (w+d+wd^2)t^{\frac{1}{2}}.$$

Define $g_c \in PGL(3,q^2)$ by $g_c : (x,y,z) \mapsto (x,y,m_c z)$. So $g_c : \overline{\mathcal{O}}_c^+ \mapsto \mathcal{O}_c^+$, and g_c fixes the two points $(0,1,0)$ and $(0,0,1)$. We show that $\overline{\mathcal{O}}_c^+$ satisfies the conclusions of the lemma. Since g_c also fixes the lines $Q, \bar{Q}$, it will follow that $\mathcal{O}_c^+$ satisfies the same conclusions.

The point $(1,t,f(t,d)) \equiv (1,y,z)$ satisfies

$$z = \frac{w^2y^4+(1+d^2)(y^3+y^2)+d^2w^2y}{y^4+w^2y^2+1} + (w+d+wd^2)y^{\frac{1}{2}}.$$

Multiply through by $y^4+w^2y^2+1$, square both sides, and then make the resulting equation homogeneous in x, y, z to get the algebraic curve $\mathcal{C}_d : H_d(x,y,z) = 0$, where

$$\begin{aligned} h_d(x,y,z) &= (z^2+(w^2+d^2+w^2d^4)xy)(y^2+wxy+x^2)^4 \\ &+wx^2y^8+wd^4x^8y^2+(1+d^4)(x^4y^6+x^6y^4). \end{aligned} \tag{6.27}$$

By construction, for all $t \in F$, $(1,t,f(t,d)) \in \mathcal{C}_d$. But in fact it is routine to check that the pointset of $\mathcal{C}_d$ is exactly the hyperoval $\overline{\mathcal{O}}_c^+$ (always $d = c/(c+w)$). So $|\mathcal{C}_d| = q+2$, and we want to prove that $h_d(x,y,z)$ is absolutely irreducible.

First compute $\frac{\partial h_d}{\partial x} = (w^2+d^2+w^2d^2)y(y^2+wxy+x^2)^4$ and $\frac{\partial h_d}{\partial y} = (w^2+d^2+w^2d^4)x(y^2+wxy+x^2)^4$ and $\frac{\partial h_d}{\partial z} = 0$. Since $w^2+x+w^2x^2$ is irreducible over $GF(4)$, and hence over F, $w^2+d^2+w^2d^4 \neq 0$ for all $d \in F$. It follows readily that the only singular point of $\mathcal{C}_d$ is $(0,0,1)$ with multiplicity 8, and $(0,0,1)$ has two tangents Q, $\bar{Q}$ in the quadratic extension of F, where the product of Q and $\bar{Q}$ is $y^2+wxy+x^2 = 0$. If one of Q, $\bar{Q}$ is a component of $\mathcal{C}_d$, so is the other, and in this case $y^2+wxy+x^2$ must divide $h_d(x,y,z)$. We now show that this is not the case. Put $y^2 = wxy+x^2$, and hence $y^4 = x^3(y+wx)$, y^6x^5y, $y^8 = wx^8+wyx^7$ into $h_d(x,y,z)$. After a little simplification, $h_d(x,y,z) = x^9(w^2(1+d^4)y+x)$, which is not the zero polynomial. This proves that neither tangent to $\mathcal{C}_d$ at $(0,0,1)$ is a component of $\mathcal{C}_d$.

Now suppose that $\mathcal{C}_d$ has components $C_1,\ldots,C_l$, $l > 1$, with $degC_i = k_i$, and suppose $(0,0,1)$ has multiplicity t_i for C_i. So $\sum k_i = 10$ and $\sum t_i = 8$. If some $t_i = k_i$, then $t_i = k_i = 1$ and C_i is a line. But in this case C_i must be tangent to C_i at $(0,0,1)$, so C_i is tangent to $\mathcal{C}_d$ at $(0,0,1)$, contradicting the previous paragraph. So the only possibility is that there are exactly two components C_1 and C_2 with $t_1 = k_1 - 1$ and $t_2 = k_2 - 1$.

First suppose C_1 and C_2 are defined over F, and $(0,0,1)$ is a $(k_i - 1)$-tuple point for C_i, $i = 1, 2$. Any tangent to C_i at $(0,0,1)$ is tangent to $\mathcal{C}_d$ and not defined over F. So any line (over F) through $(0,0,1)$ meets C_i in a second point, which implies $|C_i| \geq q+2$. Since any point of $C_1 \cap C_2$ is a singular point for $\mathcal{C}_d$, clearly $|C_1 \cup C_2| \geq 2(q+2) - 1 > q+2 = |\mathcal{C}_d|$, an impossibility.

So we may suppose that C_1 and C_2 are not defined over $F = GF(q)$, but over some $GF(q^r)$, $r > 1$. Let $id \neq \sigma \in Gal(GF(q^r)/GF(q))$. Then $C_1^\sigma = C_2$, so $degC_1 = degC_2 = 5$. By Lemma 2.24 p.55 of [Hi98], $|C_i| \leq 5^2$, so $|\mathcal{C}_d| \leq 49$. Since $q > 64$ we have a contradiction. □

Acknowledgment. The preceding proof was lifted directly from [PPP95], but the ideas of the proof originated with J. A. Thas.

For the remainder of this section write $\mathcal{O}_1^+ := \mathcal{O}_\infty^+$ and $\mathcal{O}_2^+ := \mathcal{O}_{c0}^+$. Then for $i = 1,\ 2$, let $G_i = P\Gamma L(3,q)_{\mathcal{O}_i^+}$ be the full collineation stabilizer of the hyperoval $\mathcal{O}_i^+$.

Recall that if $g \in P\Gamma L(3,q)$ has companion automorphism $\sigma \in Aut(F)$, and if $h \in F[x,y,z]$ is a homogeneous polynomial, then the *image* h^g, of h under g is defined by:

$$h^g = g^{-1} \circ h \circ \sigma,$$

where this means do g^{-1} first.

Since $g_\infty = id$, for $d = 1 \in F$, write $h_1 = h_d(x,y,z) = h_1(x,y,z)$. For $d = c_0/(c_0 + w)$ let h_2 be the image of $h_d(x,y,z)$ under g_{c0}. So in both cases, (i.e., $c = \infty; d = 1$ and $c = c_0, d = c/(c+w)$) we have the following:

$$\begin{aligned} \overline{\mathcal{O}}_c^+ &= \{(x,y,z) : h_d(x,y,z) = 0\}, \\ \mathcal{O}_c^+ &= \{g_c(x,y,z) : h_d(x,y,z) = 0\} = \{(x,y,z) : h_d(g_c^{-1}(x,y,z)) = 0\} \\ &= \{(x,y,z) : h_d^{g_c}(x,y,z) = 0\}. \end{aligned}$$

We have arranged the notation so that for $i = 1$ and $i = 2$ we have $\mathcal{O}_i^+ = \{(x,y,z) \in PG(2,q) : h_i(x,y,z) = 0\}$.

Lemma 6.9.5. *For $i = 1$, $i = 2$ we have:*

$$G_i = \{g \in P\Gamma L(3,q) : h_i^g = \lambda h_i, \lambda \in GF(q) \setminus \{0\}\}.$$

Proof. First suppose $g \in P\Gamma L(3,q)$ with $h_i^g = \lambda h_i$, $\lambda \in GF(q) \setminus \{0\}$ for $i = 1,\ 2$. Then g fixes the set of points (x,y,z) such that $h_i(x,y,z) = 0$. But these are just the points of the hyperoval $\mathcal{O}_i^+$ for $i = 1,\ 2$.

Conversely, suppose that $g \in G_i$ and suppose that $h_i^g \neq \lambda h_i$ for a nonzero λ. If $\mathcal{C}_i : h_i(x,y,z) = 0$, $i = 1,\ 2$, then the pointsets of $\mathcal{C}_i$ and ${\mathcal{C}_i}^g$ coincide. By Bezout's Theorem, it follows that the number of points of $\mathcal{C}_i \cap {\mathcal{C}_i}^g$, counted according to multiplicity, is at most 100. This is contrary to our hypothesis that $q > 64$ and $e \equiv 2 \pmod 4$, i.e., $q \geq 256$. □

The following argument is essentially the same as one in the previous section.

Since $(0,0,1)$ is the unique singular point of C_i, $i = 1, 2$, G_i must fix $(0,0,1)$ and the set $\{Q, \bar{Q}\}$ of tangents. Also, since $(0,0,1)$ is on $\mathcal{O}_i^+$, so each line through $(0,0,1)$ passes through a unique point of $\mathcal{O}_i^+$, no non-trivial element of G_i can have $(0,0,1)$ as its center. Hence G_1 and G_2 fix $(0,0,1)$ but not linewise, and so act faithfully on the quotient space $PG(2,q)/(0,0,1)$.

At this point we recall the proof of Lemma 6.6.7 to see that both G_1 and G_2 are subgroups of $C_{q+1} \rtimes C_{2e}$. We are ready to prove the main theorem of this section.

Theorem 6.9.6. *The hyperovals $\mathcal{O}_1^+$ and $\mathcal{O}_2^+$ in $PG(2,q)$, $q = 2^e$, are inequivalent for $e \equiv 2 \pmod 4$. The full collineation stabilizer of $\mathcal{O}_1^+$ in $P\Gamma L(3,q)$ is isomorphic to the semidirect product $C_5 \rtimes C_{2e}$, and the full collineation stabiliser of $\mathcal{O}_2^+$ in $P\Gamma L(3,q)$ is isomorphic to the direct product $C_5 \dot{\times} C_{e/2}$.*

Proof. For the g from Lemma 6.9.1 we have that $\langle g \rangle$ is normal in $PGL(2,q)_{l,\bar{l}}$, since $\langle g \rangle$ is characteristic in $\langle h \rangle$, which is normal in $PGL(2,q)_{l,\bar{l}}$. Since $q^2+q+1 \equiv 1 \pmod 5$, g has a unique fixed line, namely $x + y + w^2 z = 0$. Since G_1 and G_2 normalize $\langle g \rangle$, they fix $x + y + w^2 z = 0$.

Let $H_i = G_i \cap C_{q+1}$, for $i = 1, 2$. We have H_1 and H_2 fixing both tangent lines Q and $\bar{Q}$, so fixing their product $x^2 + wxy + y^2 = 0$. They also fix the line $x + y + w^2 z = 0$, so fix $x^2 + y^2 + wz^2 = 0$. Let $\mathcal{P}$ be the pencil generated by these two. The group induced on the set of conics of $\mathcal{P}$ by the stabilizer in $PGL(3,q)$ of $\mathcal{P}$ is cyclic of order $q-1$. Since $(q+1, q-1) = 1$, each H_i fixes every conic in $\mathcal{P}$. So for $i = 1, 2$, H_i fixes the conic $\mathcal{O} : xy = z^2$, so it fixes $\mathcal{O}_i^+ \cap \mathcal{O} = \mathcal{O} \cap PG(2,4)$, by Lemma 6.9.3.

Now H_i acts semiregularly on $PG(2,q) \setminus \{(0,0,1)\}$ for $i = 1, 2$. All points (except the fixed point) have orbits of length $|H_i|$. But since H_i fixes $\mathcal{O} \cap PG(2,4)$ and acts semiregularly on $\mathcal{O}$, we have $|H_i|$ dividing $|\mathcal{O} \cap PG(2,4)| = 5$. Hence $|H_i| = 1$ or 5. But $\langle g \rangle \le H_i$, so $H_i = \langle g \rangle$ for $i = 1, 2$. Hence G_1 and G_2 are subgroups of $\langle g \rangle \rtimes C_{2e}$. From 6.3.9 we know that the subgroup of H_1 induced by $\mathcal{G}_0$ has order $10e$. Hence we have $G_1 = \langle g \rangle \rtimes C_{2e}$.

As $q + 1 \equiv 0 \pmod 5$ with $q = 2^e$, then $e = 2r$ where r is odd, so $e/2$ is odd. We now show that $G_2 \le \langle g \rangle \rtimes C_{2e}$. The subgroup $C_5 \rtimes C_4$ contains all the involutions of $\langle g \rangle \rtimes C_{2e}$ (five in total) since $e/2$ is odd. These five involutions are conjugate under $\langle g \rangle$. If G_2 contains an involution, then it must contain the involution ϕ from 6.3.9. Since

$$\begin{aligned}\phi : \ & \{(1,t,f_2(t)) : t \in F\} \mapsto \\ & \{(t,1,f_2(t)) : t \in F\} = \{(1,u,uf_2(u^{-1})) : u \in F\},\end{aligned}$$

it follows that if G_2 contains an involution, then $f_2(t) = tf_2(t^{-1})$ for all $t \in F$, but this is not so. Hence 2 does not divide $|G_2|$. That is, $G_2 \le \langle g \rangle \rtimes C_{e/2}$. From Theorem 6.3.10 we have the induced group of $\mathcal{O}_2^+$ has order $5e/2$. Hence $G_2 = \langle g \rangle \rtimes C_{e/2}$. But $C_{e/2}$ has odd order and is acting on $\langle g \rangle$ where $Aut(\langle g \rangle)$

has order 4, thus it must be that the semidirect product is in fact direct. Hence $G_2 = \langle g \rangle \dot{\times} C_{e/2}$. □

6.10 Summary of Subiaco Oval Stabilizers

Let $F = GF(q)$, $q = 2^e$. Choose $\delta \in F$ so that $x^2 + \delta x + 1$ is irreducible over F. We have two "standard" ways to do this.

A. Let ζ be a primitive root of $GF(q^2) \subseteq F$. Put $\beta = \zeta^{q-1}$ and $\delta = \beta + \beta^{-1}$. This choice of δ works for any $q = 2^e$. When δ is chosen in this way, put $[a] = \beta^a + \beta^{-a}$ for rational a with denominator coprime to $q + 1$.

B. Suppose $q = 2^{2r}$, r odd (This is iff $e \equiv 2 \pmod 4$ iff $q + 1 \equiv 0 \pmod 5$.) Then put $\delta = w \in F$ where $w^2 + w + 1 = 0$.

Of course a third possibility is that $q = 2^e$ with e odd, in which case we may choose $\delta = 1$.

The basic results show that for each fixed q, each two choices of δ give equivalent q-clans. Hence we may use whichever δ is convenient. Each Subiaco hyperoval is an absolutely irreducible algebraic curve of degree 10. The complete stabilizer of the Subiaco oval is exactly the group induced by the group of the associated GQ of order (q^2, q), and this is also the complete stabilizer of the associated Subiaco hyperoval.

Refer to Section 6.5 for remarks concerning the cases $q \leq 64$. In the present section we summarize the information obtained concerning the stabilizers in each of the five cases. In each of these cases the hyperoval is given in the form

$$\mathcal{O}^+ = \{(1, t, f(t)) : t \in F\} \cup \{(0,1,0), (0,0,1)\},$$

where only the function $f(t)$ is given. $(0,0,1)$ is the nucleus of the Subiaco oval associated with the given Subiaco hyperoval.

Case 1. $q = 2^e$, e odd, $e \geq 7$. Up to projective equivalence there is only one Subiaco oval (resp., hyperoval). We may put $\delta = 1$ and choose $\mathcal{O}^+ = \mathcal{O}^+_{(1,1)}$. Then

$$f(t) = \frac{t^4 + t^3 + t^2 + t}{(t^2 + t + 1)^2} + t^{1/2}.$$

Here $P\Gamma L(3,q)_{\mathcal{O}^+}$ is a cyclic group of order $2e$ generated by $\rho : (x, y, z) \mapsto (y, x, z)$ and the automorphic collineation $(x, y, z) \mapsto (x^2, y^2, z^2)$. The homography stabilizer of $\mathcal{O}^+$ is the cyclic group $\langle \rho \rangle$ of order 2.

Case 2. $q = 2^e$, $e \equiv 0 \pmod 4$, $e \geq 8$. Up to projective equivalence there is only one subiaco oval (resp., hyperoval), and $P\Gamma L(3,q)_{\mathcal{O}^+}$ is a cyclic group of order $2e$.

We have found no simple explicit presentation of the stabilizer for any particular choice of oval. If we choose $\mathcal{O}^+ = \mathcal{O}^+_{(1,1)}$, then

$$f(t) = \frac{\delta^2 t^4 + \delta^2(1+\delta+\delta^2)(t^3+t^2) + \delta^2 t}{(t^2+\delta t+1)^2} + t^{1/2}.$$

In this case it is easy to see that the homography group is again the group $\langle \rho \rangle$, where $\rho : (x,y,z) \mapsto (y,x,z)$. However, it seems difficult to find a pleasant generator of order $2e$. On the other hand we may put $\delta = \beta + \beta^{-1}$, as explained earlier, and choose $\mathcal{O}^+ = \mathcal{O}^+_{(0,1)}$. Then

$$f(t) = \frac{(\delta^4+\delta^2)t^3 + \delta^3 t^2 + \delta^2 t}{(t^2+\delta t+1)^2} + (t/\delta)^{1/2}.$$

In this case we can say that $P\Gamma L(3,q)_{\mathcal{O}^+}$ is generated by $\hat{\psi}$ where

$$\psi = \theta(2, \frac{1}{\delta}\begin{pmatrix} [\frac{7}{10}] & [\frac{1}{5}] \\ [\frac{3}{10}] & [\frac{4}{5}] \end{pmatrix} \otimes \begin{pmatrix} \delta^{1/4} & \delta^{-1/4} \\ 0 & \delta^{-1/4} \end{pmatrix}).$$

From this it follows that $\hat{\psi}$ has the following form:

$$\hat{\psi} : (x,y,z) \mapsto$$
$$((x^2,y^2)\begin{pmatrix} [\frac{7}{5}] & [\frac{2}{5}] \\ [\frac{3}{5}] & [\frac{8}{5}] \end{pmatrix}, \delta^{3/2}z^2 + x^2\left[\frac{7}{5}\right] f\left(\frac{[\frac{2}{5}]}{[\frac{7}{5}]}\right) + y^2\left[\frac{3}{5}\right] f\left(\frac{[\frac{8}{5}]}{[\frac{3}{5}]}\right)).$$

The homography stabilizer is cyclic of order 2 generated by I_s, where $s = [\frac{2}{5}]/[\frac{3}{5}]$.

Case 3. $q = 2^e$, $e \equiv 2 \pmod 4$. Put $\delta = w$ with $w^2 + w + 1 = 0$. There are two classes of Subiaco ovals. One of them contains $\mathcal{O}_{(1,1)} = \mathcal{O}$. Then $\mathcal{O}^+$ is given by

$$f(t) = \frac{w^2(t^4+t)}{(t^2+wt+1)^2} + t^{1/2}.$$

In this case $P\Gamma L(3,q)_{\mathcal{O}^+}$ is a semidirect product $C_5 \rtimes C_{2e}$ where for a generator of C_5 we may take

$$\hat{I}_1 \circ \hat{I}_\infty : (x,y,z) \mapsto (y, x+wy, z+w^2 y),$$

and for a generator of C_{2e} we may take

$$\hat{\psi} : (x,y,z) \mapsto (x^2, x^2+w^2y^2, x^2+wz^2).$$

The homography stabilizer is dihedral of order 10 generated by the involutions $\rho = \hat{I}_1$ and $\hat{I}_\infty$.

Case 4. $q = 2^e$, $e \equiv 2 \pmod 4$, $e \not\equiv 0 \pmod 5$. Put $\delta = w$, where $w^2 + w + 1 = 0$. Here $\mathcal{O}^+ = \mathcal{O}^+_{(0,1)}$ is in the "other" class of ovals with stabilizer $C_{5e/2}$. Here

$$f(t) = \frac{t^3 + t^2 + w^2 t}{(t^2 + wt + 1)^2} + wt^{1/2}.$$

Here $P\Gamma L(3,q)_{\mathcal{O}^+}$ is cyclic of order $5e/2$ and generated by

$$\rho : (x, y, z) \mapsto (y^4, x^4 + wy^4, z^4 + wy^4).$$

The homography stabilizer is cyclic of order 5 generated by

$$\hat{I}_1 \circ \hat{I}_\infty : (x, y, z) \mapsto (y, x + wy, z + wy).$$

Case 5. $q = 2^e$, $e \equiv 10 \pmod{20}$. Choose $\delta = \beta + \beta^{-1}$, as usual. Then $\mathcal{O}^+_{(1,\delta^{1/2})}$ is in the "other" class of ovals with stabilizer $C_{5e/2}$. Here

$$f(t) = \frac{\delta^2 t^4 + \delta^5 t^3 + \delta^2 t^2 + \delta^3 t}{(t^2 + \delta t + 1)^2} + (t/\delta)^{1/2}.$$

Here $P\Gamma L(3,q)_{\mathcal{O}^+}$ is isomorphic to the direct product $C_5 \dot{\times} C_{e/2}$, and the homography stabilizer is cyclic of order 5 generated by $\hat{I}_1 \circ \hat{I}_\infty$.

Chapter 7

The Adelaide Oval Stabilizers

7.1 The Adelaide Oval

We now pick up right where we left off at the end of Section 4.8, except that **from now on we assume that** $q = 2^e$ with e even, and $m = \frac{q-1}{3} \equiv \frac{-2}{3} \pmod{q+1}$, **so we are in the Adelaide case**. The unique linear map known that stabilizes the oval $\mathcal{O}'_\alpha$ is the involution given by

$$([j], [j+1], [jm]+1) \mapsto \tag{7.1}$$

$$([j], [j+1], [jm]+1) \begin{pmatrix} 1 & [1] & 0 \\ 0 & 1 & 0 \\ 0 & 0 & 1 \end{pmatrix} = ([j], [j-1], [jm]+1).$$

The fixed points of this involution are the points of the line $x = 0$, i.e., the points $(0, y, z)$. But clearly the unique oval point on this line is the point $(0, \delta, 1)$, hence this line is a tangent line. The generator of the known stabilizer is $\hat{\theta}'$, which acts on the points of this line as $(0, y, z) \mapsto (0, \frac{y^2}{\delta}, z^2)$, from which it follows that exactly three points on this line are fixed: the oval point $(0, \delta, 1)$ and two others: $(0, 1, 0)$ and $(0, 0, 1)$. But the secant line through p'_j and p'_{-j} passes through the point $(0, 1, 0)$, implying that the nucleus must be $(0, 0, 1)$.

$$\text{Hence the line } [[j+1], [j], 0]^T \text{ is the tangent at } ([j], [j+1], [jm]+1). \tag{7.2}$$

7.2 A Polynomial Equation for the Adelaide Oval

If $q = 4$, clearly $m = 1$ and the GQ is classical. If $q = 16$, $m = 5$ and the GQ is the Subiaco GQ. Hence we assume from now on that $q \geq 64$. Since $(m, q+1) = 1$, all the ovals in the herd are projectively equivalent to the following oval:

$$\mathcal{O} = \{([j], [j+1], [jm]+1) : j \pmod{q+1}\}, \tag{7.3}$$

with known cyclic stabilizer of order $2e$ generated by

$$\theta : (x, y, z) \mapsto (x^2, \frac{x^2 + y^2}{\delta}, z^2). \tag{7.4}$$

It is easy to check that for $0 \not\equiv j \pmod{q+1}$,

$$T^2 + \delta[j]T + [2j] + \delta^2 = 0 \text{ has the two roots } [j+1] \text{ and } [j-1].$$

Hence

$$0 = tr\left(\frac{[j]^2 + \delta^2}{\delta^2[j]^2}\right) = tr\left(\frac{1}{\delta^2} + \frac{1}{[j]^2}\right)$$

for all $j \not\equiv 0 \pmod{q+1}$. But $x^2 + \delta x + 1 = 0$ has roots β and $\bar{\beta}$ which are not in F, so $tr\left(\frac{1}{\delta^2}\right) = 1$. This implies that

$$tr\left(\frac{1}{[j]}\right) = 1 \quad \text{for all } 0 \not\equiv j \pmod{q+1}. \tag{7.5}$$

Note. Since $[a] = [b]$ iff $a \equiv \pm b \pmod{q+1}$, this means that all $q/2$ elements of F with trace 1 are of the form $\frac{1}{[j]}$.

Using $[a]^3 = [3a] + [a]$ for all a mod $q+1$ (see (vi) of Lemma 3.7.1) we see that

$$[jm]^3 = \left[\frac{-2j}{3}\right]^3 = [2j] + [jm],$$

implying

$$([jm] + 1)^3 = [jm]^3 + [jm]^2 + [jm] + 1 = [2j] + ([jm] + 1)^2,$$

from which we get

$$([jm] + 1)^3 + ([jm] + 1)^2 + [2j] = 0. \tag{7.6}$$

So fix $j \not\equiv 0 \pmod{q+1}$ and put

$$G(T) = T^3 + T^2 + [2j].$$

Then $G(T) = 0$ has the root $T = [jm] + 1 = z_1$. Divide $G(T)$ by $T - z_1$ to obtain

$$G(T) = (T - z_1)(T^2 + T(1 + z_1) + (1 + z_1)z_1) = (T - z_1)H(T).$$

The quadratic factor $H(T)$ is irreducible over F, because

$$tr\left(\frac{(1+z_1)z_1}{(1+z_1)^2}\right) = tr\left(\frac{z_1}{1+z_1}\right) = tr\left(\frac{[jm]+1}{[jm]}\right) = tr\left(1 + \frac{1}{[jm]}\right) = 0 + 1 = 1,$$

by Eq. (7.5) and the fact that e is even. This proves that

$$[jm] + 1 \text{ is the unique root in } F \text{ of } T^3 + T^2 + [2j] = 0. \tag{7.7}$$

The line $[[jm]+1,\ 0,\ [j]]^T$ is a secant line to $\mathcal{O}$ containing the two oval points $([j],[j+1],[jm]+1)$ and (replacing j with $-j$) $([j],[j-1],[jm]+1)$. So if $([j],y_1,[jm]+1)$ and $([j],y_2,[jm]+1)$ are these two points, we know $y_1+y_2 = [j+1]+[j-1] = \delta[j]$, and $y_1y_2 = [j+1][j-1] = [j]^2+\delta^2$. This proves

$$([j],y,[jm]+1) \in \mathcal{O} \text{ iff } y \text{ is a root of } T^2+\delta[j]T+[j]^2+\delta^2 = 0. \tag{7.8}$$

Note the following: $[jm]+1 \neq 0$ for all j modulo $q+1$. This is clearly true for $j \equiv 0 \pmod{q+1}$. And if $j \not\equiv 0 \pmod{q+1}$, then $\frac{1}{[jm]} \neq 1$ since $tr(1) = 0$ and $tr\left(\frac{1}{[jm]}\right) = 1$. So if (x,y,z) is an arbitrary point of $\mathcal{O}$, then $z \neq 0$. If $x = 0$, then $(x,y,z) \equiv (0,\ \delta,\ 1) = \lambda(x,y,z)$ with $\lambda = z^{-1}$. If $x \neq 0$, then $(x,y,z) \equiv \left(1,\frac{y}{x},\frac{z}{x}\right)$. There is a unique $j \pmod{q+1}$, $j \not\equiv 0 \pmod{q+1}$, for which $\frac{y}{x} = \frac{[j+1]}{[j]}$. Put $\lambda = \frac{[j]}{x}$. So $\lambda(x,y,z) = (\lambda x,\ \lambda y,\ \lambda z) = ([j],\ [j+1],\ \lambda z)$ where λz $(= [jm]+1)$ is the unique root in F of $T^3+T^2+(\lambda x)^2 = 0$. This proves

$$\lambda = \frac{x^2+z^2}{z^3}, \text{ and } \lambda^2 = \frac{x^4+z^4}{z^6}. \tag{7.9}$$

By Eq. (7.8) $(\lambda y)^2+\delta(\lambda x)(\lambda y)+(\lambda x)^2+\delta^2 = 0$, so $\lambda^2(y^2+\delta xy+x^2) = \delta^2$, proving

$$\lambda^2 = \frac{\delta^2}{y^2+\delta xy+x^2}. \tag{7.10}$$

This is well defined since $y^2+\delta xy+x^2 = (y+\beta x)(y+\bar{\beta}x) \neq 0$ for nonzero $x,y \in F$.

Putting the two previous equations together, we have

Lemma 7.2.1. *If $(x,y,z) \in \mathcal{O}$, then $\delta^2z^6 = (z^4+x^4)(y^2+\delta xy+x^2)$.*

Define $G(x,y,z)$ by

$$\begin{aligned} G(x,y,z) &= \delta^2z^6+(z^4+x^4)(y^2+\delta xy+x^2) \qquad (7.11)\\ &= (x^4+z^4)y^2+(\delta x(x^4+z^4))y+\delta^2z^6+z^4x^2+x^6\\ &= \sum_{i=0}^{6} f^{(i)}(x,z)\cdot y^{6-i}, \text{ with}\\ & \quad f^{(0)} = f^{(1)} = f^{(2)} = f^{(3)} = 0;\ f^{(4)} \neq 0. \qquad (7.12) \end{aligned}$$

This shows that $(0,1,0)$ is a singular point of $C : G(x,y,z) = 0$ with multiplicity 4. The line $x = z$ is the unique tangent line of C at $(0,1,0)$ and it has multiplicity 4 at $(0,1,0)$.

It is also easy to check that

$$\frac{\partial G}{\partial x} = \delta y(x^4+z^4);\ \frac{\partial G}{\partial y} = \delta x(x^4+z^4);\ \frac{\partial G}{\partial z} = 0.$$

It now follows readily that $(0,1,0)$ is the only point (x,y,z) satisfying

$$G(x,y,z) = \frac{\partial G}{\partial x}(x,y,z) = \frac{\partial G}{\partial y}(x,y,z) = \frac{\partial G}{\partial z}(x,y,z) = 0.$$

$$\text{The unique singular point of } C : G(x,y,z) = 0 \text{ is } (0,1,0). \tag{7.13}$$

If (x_0, y_0, z_0) is a simple point of C, then the tangent line of C at this point has equation $\delta y_0(x_0^4 + z_0^4)x + \delta x_0(x_0^4 + z_0^4)y = 0$, so $y_0 x + x_0 y = 0$, and hence contains the point $(0,0,1)$ which does not belong to C.

It is of interest to check directly that the collineation θ of Eq. (7.4) does leave C invariant and fixes the point $(0,1,0)$, which is not on $\mathcal{O}$.

Theorem 7.2.2. $C = \mathcal{O} \cup \{(0,1,0)\}$.

Proof. We know that $\mathcal{O} \cup \{(0,1,0)\} \subseteq C$, so let (x,y,z) be an arbitrary point of C. If $z = 0$, then $x^4(y^2 + \delta xy + x^2) = 0$. If also $x = 0$, then we get the point $(0,1,0)$. If $x \neq 0$, then $y^2 + \delta xy + x^2$ cannot be 0 for $x, y \in F$. So (0,1,0) is the only point $(x,y,z) \in C$ with $z = 0$. Suppose $z \neq 0$. If $x = 0$, then $G(x,y,z) = 0$ implies that $y = \delta z$ and we get the point $(0, \delta, 1)$ of $\mathcal{O}$. So we may assume that $x \neq 0 \neq z$.

There is a unique $j \pmod{q+1}$ for which $\frac{y}{x} = \frac{[j+1]}{[j]}$. So

$$(x,y,z) \equiv \frac{[j]}{x}(x,y,z) = \left([j], [j+1], \frac{z[j]}{x}\right) \in C.$$

First check that $[j+1]^2 + [1][j][j+1] + [j]^2 = \delta^2$. Then from $G([j],[j+1],\frac{z[j]}{x}) = 0$, we have

$$\delta^2\left(\frac{z^6[j]^6}{x^6}\right) = \left[\left(\frac{z[j]}{x}\right)^4 + [j]^4\right]\delta^2.$$

Divide by δ^2 to get

$$\left(\frac{z[j]}{x}\right)^6 + \left(\frac{z[j]}{x}\right)^4 + [j]^4 = 0.$$

Take the square root to find that

$$\left(\frac{z[j]}{x}\right)^3 + \left(\frac{z[j]}{x}\right)^2 + [j]^2 = 0.$$

We know by Eq. (7.7) that $\frac{z[j]}{x} = [jm] + 1$. Hence $(x,y,z) \in \mathcal{O}$. □

7.3 Irreducibility of the Curve

Recall that if

$$G(x,y,z) = \delta^2 z^6 + (x^4 + z^4)(x^2 + \delta xy + y^2),$$

then

$$C = \{(x, y, z) \in PG(2, q) : G(x, y, z) = 0\}$$

has a unique singular point $P = (0, 1, 0)$ with multiplicity 4, and the line $[1, 0, 1]^T$ is the unique tangent line to C at P and it has multiplicity 4 there. Let $\hat{F}$ be an algebraic closure of F and $\hat{C} = \{(x, y, z) \in PG(2, \hat{F}) : G(x, y, z) = 0\}$. Since $x + z$ does not divide $\delta^2 z^6$, the tangent line $[1, 0, 1]^T$ is not a component of $\hat{C}$. As $\hat{C}$ has a unique singular point, each irreducible factor of $G(x, y, z)$ over $\hat{F}$ has multiplicity 1. Suppose the irreducible components of $\hat{C}$ are $\hat{C}_1, \ldots, \hat{C}_r$ for some $r > 1$, where $\deg(\hat{C}_i) = n_i$ and $\hat{C}_i$ has multiplicity m_i at P. If for some i we have $m_i = n_i$, then $m_i = n_i = 1$ and the component $\hat{C}_i$ is a line, which must therefore be a tangent to $\hat{C}_i$, and hence to $\hat{C}$, at (0,1,0). This possibility is already ruled out. Since $n_1 + \cdots + n_r = 6$ and $m_1 + \cdots + m_r = 4$ with $n_i > m_i \geq 0$ for all i, so $(n_1 - m_1) + (n_2 - m_2) + \cdots + (n_r - m_r) = 2$, it must be that $r = 2$, $n_1 = m_1 + 1$, $n_2 = m_2 + 1$, and without loss of generality we may assume that $(n_1, n_2) \in \{(1, 5), (2, 4), (3, 3)\}$. As P is the unique singular point of $\hat{C}$ it is the unique common point of $\hat{C}_1$ and $\hat{C}_2$. In particular, each of $\hat{C}_1$, $\hat{C}_2$ has the point P with multiplicity at least 1, implying $n_i \geq 2$, forcing $(n_1, n_2) = (2, 4)$ or (3,3).

Suppose that $\hat{C}_1$ and $\hat{C}_2$ are not defined over F, but rather over some extension $GF(q^s)$ with $s > 1$. Let σ be a generator of the Galois group $\mathrm{Gal}(GF(q^s)/GF(q))$. Then $\hat{C}_1^{\sigma} = \hat{C}_2$, which implies $n_1 = n_2 = 3$. Let $C_i = PG(2, F) \cap \hat{C}_i$, with $i = 1, 2$. By [Hi98], Lemma 2.24 (i), it must be that $|C_i| \leq 3^2$, so $|C| \leq 2(9) - 1 = 17$. But $|C| = q + 2$ and $q \geq 64$, so this case cannot occur. Hence $\hat{C}_1$ and $\hat{C}_2$ are defined over $F = GF(q)$.

Again let $C_i = PG(2, F) \cap \hat{C}_i$, with $i = 1, 2$. Suppose that C_1 is irreducible of degree 2 over F. Since $|C_1 \cap \mathcal{O}| \geq q$, it follows by Theorem 10.21 of [Hi98] that the unique complete arc containing $C_1 \cap \mathcal{O}$ is $\mathcal{O} \cup \{(0, 0, 1)\} = C_1 \cup \{\text{nucleus of } C_1\}$ and hence $(0, 1, 0) \in \mathcal{O}$, an impossibility. So we must have $(n_1, n_2) = (3, 3)$. This means that each component C_i is an irreducible cubic having $(0, 1, 0)$ as a unique double point (singular point with multiplicity 2) with a unique tangent at $(0, 1, 0)$.

At this point we know that C_1 and C_2 are cubic curves with a cusp at $(0, 1, 0)$ (i.e., $(0, 1, 0)$ is a double point with a unique tangent). By table 11.7, p. 260 of [Hi98], C_1 and C_2 each have $q + 1$ points with one point in common. This implies $|C| = 2q + 1$, an impossibility. This completes a proof of the following theorem.

Theorem 7.3.1. *The curve $C : G(x, y, z) = 0$ is absolutely irreducible.*

7.4 The Complete Oval Stabilizer

Lemma 7.4.1. $P\Gamma L(3, q)_{\mathcal{O}} \subseteq P\Gamma L(3, q)_{\hat{C}}$.

Proof. Let $\theta \in P\Gamma L(3, q)_{\mathcal{O}}$. Recall $C = \mathcal{O} \cup \{(0, 1, 0)\}$. Suppose $\hat{C} \neq \hat{C}^{\theta}$. Since

$\mathcal{O} \subseteq \hat{C} \cap \hat{C}^{\theta}$,

$$q+1 \leq \sum_{p \in \hat{C} \cap \hat{C}^{\theta}} m_p(\hat{C}) \cdot m_p(\hat{C}^{\theta}) \leq 6^2 = 36.$$

Hence if $q \geq 64$ then $\hat{C}^{\theta} = \hat{C}$, completing the proof. □

This means that $\theta \in P\Gamma L(3,q)_{\mathcal{O}}$ must also fix the points $(0,1,0)$ and $(0,0,1)$. Since we have a collineation fixing $\mathcal{O}$ belonging to each field automorphism, it suffices just to consider the homographies $\theta \in PGL(3,q)_{\mathcal{O}}$.

Suppose that θ is a non-identity homography

$$\theta : (x,y,z) \mapsto (x,y,z)A.$$

Since θ fixes (0,1,0) and (0,0,1) we may assume WLOG that

$$A = \begin{pmatrix} a & b & e \\ 0 & d & 0 \\ 0 & 0 & 1 \end{pmatrix}, \quad \text{with } a \cdot d \neq 0.$$

But θ must also fix $(0,\delta,1)$, the unique point of $\mathcal{O}$ on the fixed line $x=0$. Since $(0,\delta,1)A = (0,d\delta,1)$, it follows that $d=1$.

Since the point $(0,1,0)$ is fixed by θ, the secants through $(0,1,0)$ are permuted among themselves. Hence given a nonzero j modulo $q+1$ there are λ and μ in F and a unique k modulo $q+1$ for which

$$\lambda(a[j], b[j]+[j+1], e[j]+[jm]+1) = ([k],[k+1],[km]+1),$$

and

$$\mu(a[j], b[j]+[j-1], e[j]+[jm]+1) = ([k],[k-1],[km]+1).$$

Clearly $\lambda = \mu = \frac{[k]}{a[j]}$. Hence

$$\left([k], \frac{b[k]}{a} + \frac{[k][j+1]}{a[j]}, \frac{e[k]}{a} + \frac{[k]([jm]+1)}{a[j]}\right) = ([k],[k+1],[km]+1),$$

and

$$\left([k], \frac{b[k]}{a} + \frac{[k][j-1]}{a[j]}, \frac{e[k]}{a} + \frac{[k]([jm]+1)}{a[j]}\right) = ([k],[k-1],[km]+1).$$

Add together the two equations arising from the equality of the y-coordinates. The result is: $\frac{[k][j][1]}{a[j]} = [k][1]$, which implies that $a=1$. Hence

$$A = \begin{pmatrix} 1 & b & e \\ 0 & 1 & 0 \\ 0 & 0 & 1 \end{pmatrix}.$$

Since θ must also fix the line $[1,0,1]^T$, which is the tangent to $\mathcal{C}$ at the singular point $(0,1,0)$, and $(1,y,1)A = (1,b+y,e+1)$, it follows that $e = 0$.

We know that for each $y \in F$ there is a unique point $(1,y,z) \in \mathcal{O}$ for which (if $y = \frac{\lfloor j+1 \rfloor}{\lfloor j \rfloor}$, then $z = \frac{\lfloor jm \rfloor + 1}{\lfloor j \rfloor}$)

$$\delta^2 z^6 = (1+z^4)(1+\delta y+y^2).$$

Hence $(1,y,z)^\theta = (1,b+y,z)$ must satisfy

$$\delta^2 z^6 = (1+z^4)(1+\delta(b+y)+b^2+y^2),$$

implying

$$\delta^2 z^6 = (1+z^4)(1+\delta y+y^2)+(1+z^4)(\delta b+b^2).$$

As z takes on values different from 1, clearly $b = 0$ or $b = \delta$.

Of course this forces θ to be either the identity or the other linear involution that we already know stabilizes the oval $\mathcal{O}$. This concludes our proof of the following.

Theorem 7.4.2. *The complete stabilizer of the Adelaide oval is exactly the stabilizer induced by the automorphism group of the Adelaide generalized quadrangle.*

The argument used in the proof of Lemma 7.4.1 also shows that if θ fixes the hyperoval $\mathcal{O} \cup (0,0,1)$, then it also fixes the curve $\hat{C}$. But the point (0,0,1), as the intersection of the tangents to C at all points other than the singular point (0,1,0), must also be fixed. Hence θ fixes the Adelaide oval. This proves the following:

Corollary 7.4.3. *The full stabilizer of the Adelaide hyperoval is the same as the stabilizer of the Adelaide oval and is induced by the collineation group of the Adelaide GQ.*

Note. The results of this section provide an additional proof that the group $\mathcal{G}_O$ in the Adelaide case must have order $2e(q+1)$.

Chapter 8

The Payne q-Clans

8.1 The Monomial q-Clans

Suppose that $\mathcal{C} = \{A_t \equiv \begin{pmatrix} xt & yt \\ 0 & zt \end{pmatrix} : t \in F\}$ is a q-clan for which each of the functions x_t, y_t, z_t is a monomial function. In a rather remarkable paper, T. Penttila and L. Storme [PS98] show that up to the usual equivalence of q-clans, the three known examples are the only ones. Since the two non-classical families exist only for e odd, we assume throughout this chapter that e is odd. Then the three known families have the following appearance. There is some positive integer i for which $\mathcal{C} = \{A_t = \begin{pmatrix} t^{\frac{1}{2i}} & t^{\frac{1}{2}} \\ 0 & t^{\frac{2i-1}{2i}} \end{pmatrix} : t \in F\}$.

It is easy to see that when $i = 1$ the classical examples arise (because e is odd), and when $i = 2$ the FTWKB examples arise. As GQ these examples were found by W. M. Kantor [Ka80] before the notion of q-clan had appeared and supplied the initial stimulus that led to the development of q-clan geometry. When $i = 3$, the family first found by S. E. Payne [Pa85] arises. For each value of i, $1 \leq i \leq 3$, the following is an immediate corollary of Theorem 3.2.1. (Of course, for $i = 1$ and $i = 2$ we have already obtained this result.)

Corollary 8.1.1. *Suppose* $\mathcal{C} = \{A_t = \begin{pmatrix} t^{\frac{1}{2i}} & t^{\frac{1}{2}} \\ 0 & t^{\frac{2i-1}{2i}} \end{pmatrix} : t \in F\}$ *for some fixed positive integer* i. *Then* $f(t) = t^{\frac{1}{2i}}$ *and* $g(t) = t^{\frac{2i-1}{2i}}$ *satisfy* $t^{-1}g(t) = t^{-\frac{1}{2i}} = f(t^{-1})$. *So by Theorem* 3.2.1, $\theta(id, P \otimes P)$ *is an automorphism of* $GQ(\mathcal{C})$ *interchanging* $A(t)$ *and* $A(t^{-1})$. □

8.2 The Examples of Payne

Throughout this section we assume that $q = 2^e$, e odd, so $tr(1) = 1$ and $t \mapsto t^6$ is a permutation of the elements of $F = GF(q)$. Put $\mathcal{C} = \{A_t = \begin{pmatrix} t^{\frac{1}{6}} & t^{\frac{1}{2}} \\ 0 & t^{\frac{5}{6}} \end{pmatrix} : t \in F\}$. If $\mathcal{C}$ turns out to be a q-clan, by putting $i = 3$ in Cor. 8.1.1 we find a collineation ϕ

of $GQ(\mathcal{C})$ interchanging $[A(t)]$ and $[A(t^{-1})]$, $t \in \tilde{F}$. Note that $[A(1)]$ is the unique line through (∞) fixed by ϕ.

In S. E. Payne [Pa85], where these examples were first introduced, the *notion* of q-clan became clear, but the *term* was not introduced until [Pa89].

To see that $\mathcal{C}$ really is a q-clan, as well as to study the collineations of $GQ(\mathcal{C})$, it is more convenient to use an ordering of the matrices of $\mathcal{C}$ that makes $\mathcal{C}$ 3-normalized: $\mathcal{C} = \{A'_t = \left(\begin{smallmatrix} t & t^3 \\ 0 & t^5 \end{smallmatrix}\right) : t \in F\}$.

Then for $s, t \in F$, $s \neq t$, $A'_s + A'_t$ is anisotropic, since $1 + \frac{(s+t)(s^5+t^5)}{s^6+t^6} = 1 + \frac{s^4+s^3t+s^2t^2+st^3+t^4}{s^4+s^2t^2+t^4} = \frac{s^3t+st^3}{s^4+s^2t^2+t^4} = \frac{st}{s^2+st+t^2} + \left(\frac{st}{s^2+st+t^2}\right)^2$ has trace 0. Hence $\mathcal{C}$ is a q-clan.

Now let θ be any collineation of $GQ(\mathcal{C})$ fixing $(\infty), (0,0,0)$ and $[A(\infty)]$. By Theorem 2.2.1, modulo the kernel $\mathcal{N}$, we know that $\theta = \theta(\sigma, A \otimes B)$, where $\sigma \in Aut(F)$, $A \in GL(2,q)$, $B \in SL(2,q)$, $\pi : F \to F : t \mapsto \bar{t} = (\lambda t^{3\sigma} + \bar{0}^3)^{\frac{1}{3}}$, and $A'_{\bar{t}} \equiv \lambda B^{-1}(A'_t)^{\sigma} B^{-T} + A'_{\bar{0}}$. (Put $B^{-1} = \left(\begin{smallmatrix} a & b \\ c & d \end{smallmatrix}\right)$, $ad + bc = 1$).

This equivalence of matrices holds if and only if the following three conditions are satisfied:

$$\begin{array}{lrcl} \text{(i)} & \bar{t} & = & \lambda(a^2 t^{\sigma} + abt^{3\sigma} + b^2 t^{5\sigma}) + \bar{0}, \\ \text{(ii)} & \bar{t}^3 & = & \lambda t^{3\sigma} + \bar{0}^3, \\ \text{(iii)} & \bar{t}^5 & = & \lambda(c^2 t^{\sigma} + cdt^{3\sigma} + d^2 t^{5\sigma}) + \bar{0}^5. \end{array} \tag{8.1}$$

Since the expression in (ii) is the cube of that in (i), we have the following:

$$\begin{aligned} \lambda t^{3\sigma} + \bar{0}^3 &= \lambda^3(a^2 t^{\sigma} + abt^{3\sigma} + b^2 t^{5\sigma})^3 + \lambda^2(a^2 t^{\sigma} + abt^{3\sigma} + b^2 t^{5\sigma})^2 \bar{0} \\ &\quad + \lambda(a^2 t^{\sigma} + abt^{3\sigma} + b^2 t^{5\sigma})\bar{0}^2 + \bar{0}^3. \end{aligned}$$

The coefficient on $t^{15\sigma}$ is 0 on the left-hand side and $\lambda^3 b^6$ on the right-hand side. Since $\lambda \neq 0$, it must be that $b = 0$. So we have

$$\lambda t^{3\sigma} = \lambda^3 a^6 t^{3\sigma} + \lambda^2 a^4 t^{2\sigma} \bar{0} + \lambda a^2 t^{\sigma} \bar{0}^2,$$

for all $\theta \in F$. Since $b = 0$, clearly $\lambda^2 a^4 \neq 0$, so $\bar{0} = 0$. It follows that

$$\lambda a^3 = 1. \tag{8.2}$$

And $ad + bc = 1$ implies $d = a^{-1}$. Eq. (iii) now appears as: $\lambda(c^2 t^{\sigma} + cdt^{3\sigma} + d^2 t^{5\sigma}) = \bar{t}^5 = (\lambda^{\frac{1}{3}} t^{\sigma})^5 = \lambda^{\frac{5}{3}} t^{5\sigma}$. This forces $c = 0$ and $\lambda d^2 = \lambda^{\frac{5}{3}}$, or $\lambda^3 d^6 = \lambda^5$, which implies $d^3 = \lambda = a^{-3}$. So $d = a^{-1}$ as needed.

It follows that the group of special semi-linear collineations fixing $[A(\infty)]$ has order $(q-1)e$. Now return to the original ordering of the matrices in $\mathcal{C}$ and check the following with $B^{-1} = \left(\begin{smallmatrix} d^{-1} & 0 \\ 0 & d \end{smallmatrix}\right)$, $\lambda = d^3$:

$$\begin{aligned} d^3 \begin{pmatrix} d^{-1} & 0 \\ 0 & d \end{pmatrix} \begin{pmatrix} t^{\frac{\sigma}{6}} & t^{\frac{3\sigma}{6}} \\ 0 & t^{\frac{5\sigma}{6}} \end{pmatrix} \begin{pmatrix} d^{-1} & 0 \\ 0 & d \end{pmatrix} &= d^3 \begin{pmatrix} d^{-2} t^{\frac{\sigma}{6}} & t^{\frac{\sigma}{2}} \\ 0 & d^2 t^{\frac{5\sigma}{6}} \end{pmatrix} \\ &= \begin{pmatrix} (d^6 t^{\sigma})^{\frac{1}{6}} & (d^6 t^{\sigma})^{\frac{1}{2}} \\ 0 & (d^6 t^{\sigma})^{\frac{5}{6}} \end{pmatrix} = A_{d^6 t^{\sigma}}. \end{aligned}$$

It follows that $t \mapsto \bar{t} = d^6 t^\sigma$, and by Theorem 2.2.1,

$$\theta(\sigma, \begin{pmatrix} 1 & 0 \\ 0 & d^3 \end{pmatrix} \otimes \begin{pmatrix} d & 0 \\ 0 & d^{-1} \end{pmatrix}) : (\alpha, \beta, c) \mapsto$$
$$((\alpha^\sigma, \beta^\sigma)\left[\begin{pmatrix} 1 & 0 \\ 0 & d^3 \end{pmatrix} \otimes \begin{pmatrix} d & 0 \\ 0 & d^{-1} \end{pmatrix}\right], d^3 c^\sigma). \quad (8.3)$$

One consequence of these computations is that the stabilizer of $[A(\infty)]$ in $\mathcal{G}_0$ also stabilizes $[A(0)]$. Since $[A(1)]$ is the only line through (∞) fixed by ϕ, clearly $[A(\infty)]$ and $[A(1)]$ are in different orbits. And since $t \mapsto \bar{t} = d^6 t^\sigma$ shows that all lines $[A(t)]$, $0 \neq t \in F$, are in the same orbit, $\mathcal{G}_0$ has two orbits on the lines through (∞), one of which is $\{[A(\infty)], [A(0)]\}$. And $|\mathcal{G}_0/\mathcal{N}| = 2(q-1)e$.

We now consider the action of $\mathcal{G}_0$ on the associated ovals and subquadrangles, still using the $\frac{1}{2}$-normalized form of the q-clan. Clearly the involution $\phi = \theta(id, P \otimes P)$ interchanges $\mathcal{R}_{(1,0)}$ and $\mathcal{R}_{(0,1)}$. Also, $\theta = \theta(\sigma, \left(\begin{smallmatrix} 1 & 0 \\ 0 & d^3 \end{smallmatrix}\right) \otimes \left(\begin{smallmatrix} d & 0 \\ 0 & d^{-1} \end{smallmatrix}\right))$ maps $\mathcal{R}_{(1,t)}$ to $\mathcal{R}_{(1,t^\sigma d^{-2})}$.

Hence it suffices to consider just two cases.

Case (i). $\alpha = (1,0)$. Since $(1,0)^\sigma \left(\begin{smallmatrix} d & 0 \\ 0 & d^{-1} \end{smallmatrix}\right) = d(1,0)$, we have that θ stabilises $\mathcal{R}_{(1,0)}$ (with $\lambda = d$ in the notation of Theorem 2.8.1). Here

$$\mathcal{O}_{(1,0)} = \{(1, t, t^{\frac{1}{6}}) : t \in F\} \cup \{(0,1,0)\}.$$

In the notation of Theorem 2.8.1 we have $C = 0 = E$. Hence

$$\hat{\theta} : (1, t, t^{\frac{1}{6}}) \mapsto (d^2(1, t^\sigma)\begin{pmatrix} 1 & 0 \\ 0 & d^6 \end{pmatrix}, d^3 t^{\frac{\sigma}{6}}) = (d^2, d^8 t^\sigma, d^3 t^{\frac{\sigma}{6}}) \equiv (1, d^6 t^\sigma, d t^{\frac{\sigma}{6}}).$$

As a collineation of $PG(2,q)$,

$$\hat{\theta} : (x_0, x_1, x_2) \mapsto (x_0^\sigma, x_1^\sigma, x_2^\sigma)\begin{pmatrix} 1 & 0 & 0 \\ 0 & d^6 & 0 \\ 0 & 0 & d \end{pmatrix}.$$

This gives a group of order $(q-1)e$ stabilizing

$$\mathcal{O}_{(1,0)} = \{(x_0, x_1, x_2) : x_2^6 = x_0^5 x_1\}$$

which is proved to be the complete oval stabiliser in [Pa86].

Case (ii). $\alpha = (1,1)$. Here $g((1,1),t) = t^\delta$, where $\delta : t \mapsto t^\delta = t^{\frac{1}{6}} + t^{\frac{3}{6}} + t^{\frac{5}{6}}$, then

$$\mathcal{O}_{(1,1)} = \{(1, t, t^\delta) : t \in F\} \cup \{(0,1,0)\}.$$

Since $(1,1)P = (1,1)$, ϕ stabilizes $\mathcal{R}_{(1,1)}$ with $\lambda = 1$. And

$$\hat{\phi} = \hat{\theta}(id, P \otimes P) : (1, t, t^\delta) \mapsto ((1,t)P^{(2)}, t^\delta) = (t, 1, t^\delta).$$

To find the induced stabilizer of $\mathcal{O}_{(1,1)}$ it remains only to observe that $\theta(\sigma, \left(\begin{smallmatrix} 1 & 0 \\ 0 & d^3 \end{smallmatrix}\right) \otimes \left(\begin{smallmatrix} d & 0 \\ 0 & d^{-1} \end{smallmatrix}\right))$ stabilizes $\mathcal{R}_{(1,1)}$ if and only if $d = 1$. Then

$$\hat{\theta} = \hat{\theta}(\sigma, I \otimes I) : (x_0, x_1, x_2) \mapsto (x_0^\sigma, x_1^\sigma, x_2^\sigma).$$

This gives an induced stabilizer of order $2e$, which is easily seen to be cyclic and by [TPG88] must be the complete stabilizer of the oval.

8.3 The Complete Payne Oval Stabilizers

The Stabilizer of $\mathcal{O}_{(1,0)}$

The oval $\mathcal{O}_{(1,0)} = \{(1, t, t^{\frac{1}{6}}) : t \in F\} \cup \{(0, 1, 0)\} \equiv \{(1, t, t^6) : t \in F\} \cup \{(0, 1, 0)\}$ is known to be the Segre oval (it was discovered by Segre [Se62] in 1962). In [Hi98], p. 190, we find that in $PG(2, 8)$ every hyperoval is regular, i.e., it contains a conic. In particular, the Segre oval involved in this case is a "pointed conic[1]." So let $q \geq 2^5 = 32$. From [Pa86] we know that the complete subgroup of $P\Gamma O(3, q)$ stabilising $\mathcal{O}_{(1,0)}$ has order $(q - 1)e$. To obtain such a result, Theorem 8.3.1 is used, which holds for any hyperoval $\mathcal{O}^+(\alpha) = \{A, B\} \cup \{(1, c, c^\alpha) \colon c \in F\}$, where $A = (0, 1, 0)$, $B = (0, 0, 1)$ and α is a permutation of the elements of F fixing 0 and 1 and satisfying

$$(*) : \ (c_0^\alpha - c_1^\alpha)(c_0 - c_2) \neq (c_0^\alpha - c_2^\alpha)(c_0 - c_1)$$

with c_0, c_1, $c_2 \in F$. For each integer k, the map $\alpha \colon x \mapsto x^k$ is a permutation of the elements of F fixing 0 and 1 if and only if $gcd(k, q - 1) = 1$. For such an α, write k in place of α, put $M = \{k \colon 1 \leq k \leq q - 1,\ gcd(k, q - 1) = 1\}$, $\mathcal{I} = \{k \in M$ such that k satisfies condition $(*)\}$ and $\mathcal{A} = \{\alpha \colon \alpha$ satisfies $(*)$ and α is additive$\}$ $=$ $\{k \in M \colon k = 2^i,\ 1 \leq i \leq e,\ gcd(i, e) = 1\}$.

Theorem 8.3.1 ([Pa86], Cor. IV.1). *Let $\alpha \in \mathcal{I}$, with α, $\alpha/(\alpha - 1)$, $1 - \alpha \notin \mathcal{A}$. Let G be the group of collineations of $PG(2, q)$ stabilising $\mathcal{O}^*(\alpha) = \{A, B, C\} \cup \mathcal{O}^0$ where $C = (1, 0, 0)$ and $\mathcal{O}^0 = \mathcal{O}^* \setminus \{A, B, C\}$. Then*

(i) *$G_A = G_B = G_C$ and G_A acts transitively on $\mathcal{O}^0$ so that its one-point stabiliser in $\mathcal{O}^0$ has no further one-point orbit in $\mathcal{O}^0$.*

(ii) *$\{A, B, C\}$ is invariant under G; $\mathcal{O}^0$ is an orbit of G.*

(iii) *$G = G_A$ unless $\alpha^2 - \alpha + 1 \equiv 0 \ (mod\, q - 1)$, in which cases the map $(x, y, z) \mapsto (z, x, y)$ induces a collineation of $PG(2, q)$ fixing $\mathcal{O}^*(\alpha)$, having the cycle (A, B, C) and mapping $(1, c, c^\alpha)$ to $(1, d, d^\alpha)$, with $d = c^{-\alpha}$.*

(iv) *G_A is generated by the collineations $\theta \colon (x, y, z) \mapsto (x^\sigma, y^\sigma, z^\sigma)$, $\sigma \in Aut(F)$ and $\varphi \colon (x, y, z) \mapsto (x, dy, d^\alpha z)$, $0 \neq d \in F$.*

[1]For q even, $q > 4$, if $\mathcal{C}$ is a conic with nucleus N and $P \in \mathcal{C}$, then $(\mathcal{C} \cup \{N\}) \setminus \{P\}$ is an oval inequivalent to a conic, called a *pointed conic*.

Clearly, in our case $\alpha = 6$ and $\mathcal{O}(\alpha) = \mathcal{O}^*_{(1,0)}$. Moreover, the complete subgroup of $P\Gamma O(3, q)$ stabilizing the hyperoval $\mathcal{O}^*_{(1,0)}$ containing $\mathcal{O}_{(1,0)}$ also has order $(q-1)e$ except in the one case $q = 2^5$. In that case $6^2 - 6 + 1 = 31 \equiv 0 \pmod{q-1}$, so there is an extra factor of 3 stabilizing the hyperoval ([Pa86]). But in all cases with $q \geq 2^5$, $\mathcal{G}_0$ induces the complete stabilizer of the oval $\mathcal{O}_{(1,0)}$.

The Stabilizer of $\mathcal{O}_{(1,1)}$

For this section we refer to the paper [TPG88]. Recall that

$$\mathcal{O}_{(1,1)} = \{(1, t, t^\delta) : t \in F\} \cup \{(0, 1, 0)\}$$

where $\delta : t \mapsto t^\delta = t^{\frac{1}{6}} + t^{\frac{3}{6}} + t^{\frac{5}{6}}$ and whose nucleus is $(0, 0, 1)$. Since δ commutes with each automorphism of the field, $\hat{\theta} = \hat{\theta}(\sigma, I \otimes I) : (x, y, z) \mapsto (x^\sigma, y^\sigma, z^\sigma)$ generates a group of order e of collineations of $PG(2, q)$ leaving invariant the oval $\mathcal{O}_{(1,1)}$. It is easy to see that $(x^{-1})^\delta = \frac{x^\delta}{x}$, from which it follows that $\hat{\phi} : (x, y, z) \mapsto (y, x, z)$ is a projectivity of $PG(2, q)$ fixing the points $(1, 1, 1)$ and $(0, 0, 1)$ and interchanging $(1, 0, 0)$ with $(0, 1, 0)$ and $(1, c, c^\delta)$ with $(1, c^{-1}, (c^{-1})^\delta)$, $c \neq 0$. Hence $\hat{\phi}$ leaves invariant the hyperoval $\mathcal{O}^*{}_{(1,1)} = \mathcal{O}_{(1,1)} \cup \{(0, 0, 1)\}$. The group $G = \langle \hat{\theta}, \hat{\phi} \rangle$ has order $2e$ and we will show that it is the full group G^* of collineations of $PG(2, q)$ leaving invariant the hyperoval $\mathcal{O}^*{}_{(1,1)}$. The orbits of G on $\mathcal{O}^*{}_{(1,1)}$ are the two singletons $\{(0, 0, 1)\}$ and $\{(1, 1, 1)\}$, the set $\{(1, 0, 0), (0, 1, 0)\}$ and the sets of size $2d$, where $1 < d$ and d divides e.

An algebraic curve of degree 6

It can be easily verified that $(x^\delta)^6 = x(x^\delta)^4 + x^5 + x$. Take square roots, put $u = x^\delta$ and define $f_x(u) = u^3 + x^{\frac{1}{2}} u^2 + x^{\frac{5}{2}} + x^{\frac{1}{2}}$ which is the same as $f_x(u) = (u - x^\delta)(u^2 + (x^{\frac{1}{6}} + x^{\frac{5}{6}})u + x^\delta(x^{\frac{1}{6}} + x^{\frac{5}{6}}))$.

Theorem 8.3.2. *If $x \neq 1 \in F$, then x^δ is the unique root in F of $f_x(u) = 0$. If $x = 1$ then both $1^\delta = 1$ or $u = 0$ are roots of $f_1(u) = 0$.*

Proof. For $x = 0$, $f_0(u) = u^3 = 0$ having $u = 0^\delta = 0$ as root. For $x = 1$, $f_1(u) = u^2(u + 1) = 0$ having as roots both $u = 1^\delta = 1$ and $u = 0$. Recall the elements of F are partitioned into two sets: $C_0 = \{z^2 + z : z \in F\}$ and its other additive coset C_1, where since e is odd $C_1 = C_0 + 1$. Then $u^2 + au + b$ is irreducible over F if and only if $b/a^2 \in C_1$ (see [PT84]). So for $0 \neq x \neq 1$, $u^2 + (x^{\frac{1}{6}} + x^{\frac{5}{6}})u + x^\delta(x^{\frac{1}{6}} + x^{\frac{5}{6}})$ is irreducible over F iff $x^\delta(x^{\frac{1}{6}} + x^{\frac{5}{6}})/(x^{\frac{1}{6}} + x^{\frac{5}{6}})^2 = 1 + x^{1/3}/(1 + x^{1/3})^2 \in C_1$, which is the case since $1 \in C_1$ and $x^{1/3}/(1 + x^{1/3})^2 = A^2 + A \in C_0$, with $A = (1 + x^{1/3})^{-1}$. □

Writing homogeneous coordinates for the points of $\mathcal{O}_{(1,1)}$, we find

$$\begin{aligned}\mathcal{O}_{(1,1)} &= \{(x,y,z) : x = z = 0 \neq y \text{ or } x \neq 0 \text{ and } (y/x)^{\delta} = z/x\} \\ &= \{(0,1,0),(1,1,1)\} \cup \{(x,y,z) : x \neq y \neq 0 \text{ and } \\ &\quad \left(\frac{z}{x}\right)^6 + \left(\frac{y}{x}\right)\left(\frac{z}{x}\right)^4 + \left(\frac{y}{x}\right)^5 + \frac{y}{x} = 0\} \\ &= \{(0,1,0),(1,1,1)\} \cup \{(x,y,z) : x \neq y \neq 0 \text{ and } z^6 = xy(x+y+z)^4\}. \quad (8.4)\end{aligned}$$

Putting $x = 0$ in $z^6 = xy(x+y+z)^4$ allows only the point $(0,1,0)$ while $x = y \neq 0$ does allow either the point $(1,1,1)$ or also the 'extraneous' point $(1,1,0)$. Define the algebraic curve Γ in $PG(2,q)$ by

$$\Gamma = \{(x,y,z) \in PG(2,q) : z^6 = xy(x+y+z)^4\}.$$

All this proves the following

Theorem 8.3.3. $\Gamma = \mathcal{O}_{(1,1)} \cup \{(1,1,0)\}$.

The special point $(1,1,0)$

The following results point out the special role played by the point $(1,1,0)$. For any $c \in F \setminus \{0,1\}$, the line joining the points $(1,c,c^{\delta})$ and $(1,c^{-1},(c^{-1})^{\delta})$ of $\mathcal{O}_{(1,1)}$ contains the point $(1,1,0)$. Recall that in $PG(2,q)$, q even, the three diagonal points (i.e., the points obtained pairwise intersecting the lines joining the vertices of a quadrangle) are on a line.

Proposition 8.3.4. *If the distinct lines p_1p_2 and q_1q_2, where $p_1, p_2, q_1, q_2 \in \mathcal{O}_{(1,1)}$ contain the point $(1,1,0)$, then the line joining the diagonal points of the complete quadrangle $p_1p_2q_1q_2$ is always the line $[1,1,0]$.*

Proposition 8.3.5. *Let p_1p_2, q_1q_2, r_1r_2, where $p_1, p_2, q_1, q_2, r_1, r_2, \in \mathcal{O}_{(1,1)}$ be distinct lines containing the point $(1,1,0)$. Then $p_1, p_2, q_1, q_2, r_1, r_2$ belong to an irreducible conic C. Moreover, the line $[1,1,0]$ is the tangent to C from $(1,1,0)$.*

Proposition 8.3.6. *Let $(1,1,0)$ be on the line p_1p_2 with p_1, p_2 points of $\mathcal{O}_{(1,1)}$. Then the line joining the diagonal points of p_1p_2rs with $r = (0,0,1)$ and $s = (1,1,1)$, is external to $\mathcal{O}_{(1,1)}$. In this way there arise all $q/2$ lines through $(1,1,0)$ having no point in common with $\mathcal{O}_{(1,1)}$.*

The Complete (Hyper)Oval Stabilizer for $q > 32$

Our aim is to show that for $e > 5$ the complete group G^* of collineations of $PG(2,q)$ leaving invariant the hyperoval is exactly the group G spanned by $\hat{\phi}$ and $\hat{\theta}$. The case $e = 5$ is more difficult and, even though it yields the same results, we prefer to treat it separately. Since $\sigma \in G^*$, it suffices to determine all linear collineations of $PG(2,q)$ leaving $\mathcal{O}^*$ invariant.

Theorem 8.3.7. *Let α be a linear collineation in G^*. If $q > 32$ then $\alpha \in \{id, \hat{\phi}\}$.*

Proof. The linear collineation α transforms the algebraic curve Γ into an algebraic curve Γ' of degree 6. The point $(1,1,0)$ is the only singular point of Γ with multiplicity 4, hence $(1,1,0)^\alpha$ is the unique singular point of Γ'. Suppose $\Gamma \neq \Gamma'$. Since Γ and Γ' are irreducible, by Bezout's theorem we have $|\Gamma \cap \Gamma'| \leq 36$. As $\mathcal{O}_{(1,1)} \cap \mathcal{O}^\alpha_{(1,1)} \subset \Gamma \cap \Gamma'$, we have $q \leq |\Gamma \cap \Gamma'|$. If $q > 32$ we get a contradiction, hence $\Gamma = \Gamma'$ and $(1,1,0)^\alpha = (1,1,0)$. The tangents of Γ at the simple points of Γ concur at $(0,0,1)$, so we have $(0,0,1)^\alpha = (0,0,1)$. Let L be a line through $(0,0,1)$. If $L \cap \mathcal{O}_{(1,1)} = P$, then the intersection multiplicity of L and Γ at P is exactly 6 iff L is $x = 0$ or $y = 0$, in which case P is $(0,1,0)$ or $(1,0,0)$. Hence, with $[a,b,c]$ denoting the line with equation $aX + bY + cZ = 0$, we have the following: $\mathcal{O}^\alpha_{(1,1)} = \mathcal{O}_{(1,1)}$; $(1,1,0)^\alpha = (1,1,0)$; $(0,0,1)^\alpha = (0,0,1)$; $[1,1,0]^\alpha = [1,1,0]$; $(1,1,1)^\alpha = \{\mathcal{O}_{(1,1)} \cap [1,1,0]\}^\alpha = \mathcal{O}^\alpha_{(1,1)} \cap [1,1,0]^\alpha = \mathcal{O}_{(1,1)} \cap [1,1,0] = (1,1,1)$; $\{(0,1,0),(1,0,0)\}^\alpha = \{(0,1,0),(1,0,0)\}$; $[0,0,1]^\alpha = [0,0,1]$. Since α is linear, it is easy to check that $\alpha \in \{id, \hat{\phi}\}$. □

It is clear that the complete hyperoval stabilizer coincides with the complete oval stabilizer, hence

Corollary 8.3.8. *The complete Payne oval stabilizer for $q > 32$ is $\langle \hat{\phi}, \hat{\theta} \rangle$ of order $2e$.*

The Complete Payne (Hyper)Oval Stabilizer for q = 32

For $q = 32$ we refer to the paper [TPG88] and to the paper [OPP91]. In the first paper the authors show that the following holds.

Lemma 8.3.9. *Let α be a linear collineation in G^*. Then*

(i) *if $q = 32$ and $(1,1,0)^\alpha = (1,1,0)$, then $\alpha \in \{id, \hat{\phi}\}$;*

(ii) *if $q = 32$ and $(0,0,1)^\alpha = (0,0,1)$, then $\alpha \in \{id, \hat{\phi}\}$.*

Proof. Suppose $q = 32$ and $(1,1,0)^\alpha = (1,1,0)$. The tangents of Γ at its simple points (i.e., the points of $\mathcal{O}_{(1,1)}$) concur at $(0,0,1)$. Consequently, the tangents of Γ' at the points of $\mathcal{O}^\alpha_{(1,1)} = \mathcal{O}_{(1,1)}$ concur at $(0,0,1)$. Therefore the tangents of Γ and Γ' at the points of $\mathcal{O}_{(1,1)}$ coincide. This means that, if $\Gamma \neq \Gamma'$ and considering the intersection multiplicities of points of $\Gamma \cap \Gamma'$, the points of $\mathcal{O}_{(1,1)}$ account for at least $2(q+1) = 66$ common points, contradicting Bezout's theorem. Hence we have $\Gamma = \Gamma'$ and also $(1,1,0)^\alpha = (1,1,0)$.

Suppose now that $q = 32$ and $(0,0,1)^\alpha = (0,0,1)$, so the singular points of multiplicities 4 of Γ and Γ' coincide. Assume $\Gamma \neq \Gamma'$. Then $(1,1,0)$ accounts for at least $4 \times 4 = 16$ common points of Γ and Γ'. Since $32 + 16 > 36$, we again have a contradiction by Bezout's theorem. So in both cases we have $\Gamma = \Gamma'$. Now, just rephrasing the last part of Theorem 8.3.7, the lemma follows. □

Having established that, they prove (also using the help of the computer) that if $q = 32$, then $(1,1,0)^\alpha = (1,1,0)$ for any linear collineation $\alpha \in G^*$ (we refer to the paper [TPG88] for the details), from which it follows that $G^* = \langle \hat{\phi}, \hat{\theta} \rangle$.

In the paper [OPP91] the authors use a more algebraic approach to the problem yielding to a computer-free proof. Let H denote the homography stabilizer of $\mathcal{O}^*$ and as usual let G^* be the collineation stabiliser of $\mathcal{O}^*$ in $PG(2,q)$. Clearly, $|G^*| = e|H| = 5|H|$. They find that $|H| \equiv 2 \pmod 5$, and since $|H| = 2^i$ for some $0 \le i \le 6$, the only possible values for $|H|$ are 2 and 32. The case $|H| = 32$ leads to a contradiction, so $|H| = 2$ and since $|G : H| = 5$ we have $|G| = 10$ and so $G^* = \langle \hat{\phi}, \hat{\theta} \rangle$. Hence we get the analogue of Corollary 8.3.8 also for $q = 32$.

Theorem 8.3.10. *The complete Payne oval stabilizer for* $q = 32$ *is* $\langle \hat{\phi}, \hat{\theta} \rangle$ *of order* $2e$.

Chapter 9

Other Good Stuff

This chapter contains very brief introductions to a few important topics and results related to q-clan geometries. These results were not needed for the general development of the material in the preceding chapters, but in many cases they provide resolutions of natural questions arising from the study of the known examples and show the special importance of these examples in the general theory.

9.1 Spreads and Ovoids

Let $\mathcal{S}$ be a GQ with parameters (s,t), $s \geq 1$, $t \geq 1$. A *spread* of $\mathcal{S}$ is a set $\mathcal{M}$ of lines that partition the points of $\mathcal{S}$. Dually, an *ovoid* of $\mathcal{S}$ is a set $\mathcal{O}$ of points of $\mathcal{S}$ such that each line of $\mathcal{S}$ is incident with a unique point of $\mathcal{O}$. It is easy to see that a spread must have $1+st$ lines and an ovoid must have $1+st$ points. For example, if a GQ $\mathcal{S}'$ of order q is contained as a subquadrangle in a GQ $\mathcal{S}$ with order (q^2, q), then each point X of a line l exterior to $\mathcal{S}'$ is on a unique line of $\mathcal{S}'$. Hence the q^2+1 lines of $\mathcal{S}'$ that meet l form a spread of $\mathcal{S}'$ said to be subtended by l. Spreads and ovoids of GQ have been studied a great deal and have a wide variety of connections with other geometric objects. For a general reference see J. A. Thas and S. E. Payne [TP94]. For $q = 2^e$ see especially [BOPPR1] and [BOPPR2]. In this section we give a very brief introduction to the material contained in these latter two papers.

Let $\mathcal{O}_1$ and $\mathcal{O}_2$ be ovals of $PG(2,q)$ and let P be a point of $PG(2,q)$ not on either oval and different from their nuclei. Then $\mathcal{O}_1$ and $\mathcal{O}_2$ are *compatible at* P provided they have the same nucleus, they have a point Q in common, the line $\langle P, Q\rangle$ is tangent to both ovals, and every secant line to $\mathcal{O}_1$ on P is external to $\mathcal{O}_2$. As a consequence, every line on P which is secant to $\mathcal{O}_2$ is external to $\mathcal{O}_1$.

Let f be an o-polynomial over $GF(q)$, so in particular $f(1) = 1$. A *generalized f-fan* in $PG(2,q)$ is a family $\mathcal{F} = \{\mathcal{O}_s : s \in GF(q)\}$ of ovals in $PG(2,q)$ such that:

- $\mathcal{O}_s$ has nucleus $(0,1,0)$ for each $s \in GF(q)$.

- $\mathcal{O}_s \cap \mathcal{O}_t = (0,0,1)$ for all distinct $s,\ t \in GF(q)$.
- $\mathcal{O}_s$ and $\mathcal{O}_t$ are compatible at the point $P_{st} = (0, 1, \frac{f(s)+f(t)}{s+t})$ for all $s \neq t$.

The line $[1,0,0]^T$ is called the *common tangent* of $\mathcal{F}$. If $\mathcal{F}$ is a generalized f-fan for some o-polynomial f where f is not stated explicitly, we say simply that $\mathcal{F}$ is a generalized fan. An *augmented fan* in $PG(2,q)$ is a family $\mathcal{A}$ of q ovals in $PG(2,q)$ indexed by distinct points of $PG(2,q)$ such that:

(1) Each oval of $\mathcal{A}$ has the same nucleus N;

(2) The ovals of $\mathcal{A}$ intersect pairwise in a fixed point P;

(3) The q indexing points together with P form an oval $\mathcal{O}$ with nucleus N;

(4) $\mathcal{O}_Q$ and $\mathcal{O}_R$ in $\mathcal{A}$ indexed by Q and R respectively, are compatible at $\langle Q, R\rangle \cap \langle P, N\rangle$.

The oval $\mathcal{O}$ is called the *augmenting oval* and the remaining ovals are called the *ovals of the fan.* A generalized f-fan gives rise to an augmented fan with $\mathcal{O} = D(f) = \{(1,t,f(t)) : t \in GF(q)\} \cup \{(0,0,1)\}$, $P = (0,0,1)$, $N = (0,1,0)$, $\mathcal{O}_t$ indexed by $(1,t,f(t))$, and conversely.

The main interest in augmented fans derives from the fact that there is a correspondence between augmented fans with augmenting oval $\mathcal{O}$ and spreads of $T_2(\mathcal{O})$. The definition of $T_2(\mathcal{O})$ is given in the Prolegomena.

Theorem 9.1.1. *([BOPPR1]) Let $\mathcal{O}$ be an oval of $PG(2,q)$ with nucleus N, and let Γ be a spread of $T_2(\mathcal{O})$. Then Γ contains a line of $T_2(\mathcal{O})$ of type (b), i.e., a point P of $\mathcal{O}$. Let π be a plane of $PG(3,q)$ meeting π_∞ in the tangent line $l = \langle P, N\rangle$ to $\mathcal{O}$ at P. Each point X of $\mathcal{O}\setminus\{P\}$ is on q lines of Γ, and these lines, together with the line joining X and P, project an oval $\mathcal{O}_X$ of π with nucleus N. Moreover, if $Y \in \mathcal{O}\setminus\{P,X\}$, $\mathcal{O}_X$ and $\mathcal{O}_Y$ are compatible at $\langle X,Y\rangle \cap l$. Taking the image of $\mathcal{O}$ under a collineation of $PG(3,q)$ mapping π_∞ to π as the augmenting oval, gives an augmented fan. Conversely, given an augmented fan with augmenting oval $\mathcal{O}$, construct $T_2(\mathcal{O})$, choose a plane π meeting π_∞ in the distinguished tangent line l of $\mathcal{O}$, take the image of the augmented fan in π_∞ under a collineation of $PG(3,q)$ mapping π_∞ to π and fixing l pointwise, and for each point X of $\mathcal{O}\setminus l$, join X to all points of the image of $\mathcal{O}_X$ not on l, to obtain q^2 lines of type (a) of $T_2(\mathcal{O})$. Then adding the line $P = l \cap \mathcal{O}$ of type (b), there arises a spread Γ of $T_2(\mathcal{O})$.*

Probably the main result of interest to mention here is the following

Corollary 9.1.2. *Every augmented fan of conics arises from a normalized flock of the quadratic cone, and conversely, every flock of the quadratic cone arises from an augmented fan of conics.*

This last result in a slightly different form and with a geometric proof first appeared in [Th01b]. The use of generalized f-fans to represent spreads of $T_2(\mathcal{D}(\mathcal{O}))$ generalizes the use of fans to represent ovoids of $PG(3,q)$ (as in [Pe99] and [Gl98]), in the following sense. An ovoid Ω of $PG(3,q)$ is an ovoid of the GQ $W(q)$ constructed from the symplectic polarity defined by Ω. Since $W(q)$ is self-dual for q even, the ovoid Ω of $W(q)$ gives rise to a spread of $W(q)$. For q even, the GQ $W(q)$ is isomorphic to the GQ $T_2(\mathcal{O})$ where $\mathcal{O}$ is a conic. Hence the spread of $W(q)$ gives rise to a spread of $T_2(\mathcal{O})$ where $\mathcal{O}$ is an oval given by the o-polynomial $f(t) = t^2$.

The preceding approach to studying spreads of $T_2(\mathcal{O})$ yields a great deal more. We have not mentioned here the connection between spreads of $T_2(\mathcal{O})$ with $\mathcal{O}$ a translation oval and α-flocks (see later for the definition of α-flock). (See [BOPPR1], [BOPPR2] and [C98].) In fact the above approach leads to the following characterization of those ovals that can belong to a herd.

Theorem 9.1.3. *Let $\mathcal{O}$ be an oval of $PG(2,q)$, q even, with nucleus N. Then $\mathcal{O}$ is projectively equivalent to an oval contained in a herd if and only if the GQ $T_2(\mathcal{O})$ has a spread consisting of an element of $\mathcal{O}$ and q^2 lines of $PG(3,q)$ contained in q quadratic cones, where the nuclear line of each cone contains N.*

Using the general theory alluded to in this section all spreads of $T_2(\mathcal{O})$ for fields of order at most 32 have been determined (see [BOPPR1]), with the consequence that many $T_2(\mathcal{O})$ of order $q = 32$ are known not to be contained as a subquadrangle of any GQ of order (q^2, q) or (q, q^2). It is worthwhile to mention here the rather remarkable result obtained by Brown [Br00] regarding ovoids of $PG(3,q)$ with a conic plane section

Theorem 9.1.4. *An ovoid of $PG(3,q)$ having a conic as plane section is an elliptic quadric.*

9.2 The Geometric Construction of J. A. Thas

The problem of finding a geometric construction of a flock GQ was open for quite some time. As the years passed after N. Knarr [Kn92] found such a construction for q odd, many researchers came to believe that no such construction could exist for q even. Hence it was quite surprising when J. A. Thas presented his construction in [Th99], with refinements of related material in [Th98] and [Th01a]. This construction represents a major resolution of a long open problem, and no treatment of flock GQ could be considered complete without it. We just give the construction without any attempt to sketch the long difficult proof.

Let $\mathcal{C}$ be a quadratic cone with vertex x of $PG(3,q)$. Further, let y be a point of $\mathcal{C} \setminus \{x\}$ and let ζ be a plane of $PG(3,q)$ not containing y. Project $\mathcal{C} \setminus \{y\}$ from y onto ζ. Let τ be the tangent plane of $\mathcal{C}$ at the line xy and let $\tau \cap \zeta = T$. Then with the q^2 points of $\mathcal{C} \setminus xy$ there correspond the q^2 points of the affine plane $\zeta \setminus T = \zeta'$; with any point of $xy \setminus \{y\}$ there corresponds the intersection ∞ of xy and ζ; with the generators of $\mathcal{C}$ distinct from xy there correspond the lines of ζ distinct from T containing ∞; with the (nonsingular) conics on $\mathcal{C}$ not passing through y there correspond the $q^2(q-1)$ (nonsingular) conics of ζ which are tangent to T at ∞.

Let $F = \{C_1^*, C_2^*, \ldots, C_q^*\}$ be a flock of the cone $\mathcal{C}$. Now consider the set $\tilde{F} = \{C_1, C_2, \ldots, C_{q-1}, N\}$ consisting of the $q-1$ nonsingular conics C_1, C_2, ..., C_{q-1} and the line N of ζ, which is obtained by projecting the elements of F from y onto ζ. So C_1, C_2, ..., C_{q-1} are conics which are mutually tangent at ∞ (with common tangent line T) and N is a line of ζ not containing ∞.

Now we consider planes $\pi_\infty \neq \zeta$ and $\mu \neq \zeta$ of $PG(3,q)$, respectively containing T and N. In μ we consider a point r with $r \notin \zeta \cap \pi_\infty$. Next, let O_i be the nonsingular quadric which contains C_i, which is tangent to π_∞ at ∞ and which is tangent to μ at r, with $i = 1, 2, \ldots, q-1$. As $C_i \cap N = \emptyset$, the quadric O_i is elliptic, $1 \leq i \leq q-1$.

Next let $\mathcal{S}$ be the following incidence structure.

Points of $\mathcal{S}$

(a) The $q^3(q-1)$ nonsingular elliptic quadrics O containing $O_i \cap \pi_\infty = L_\infty^{(i)} \cup M_\infty^{(i)}$ (over $GF(q^2)$) such that the intersection multiplicity of O_i and O at ∞ is at least three (these are O_i, the nonsingular elliptic quadrics $O \neq O_i$ containing $L_\infty^{(i)} \cup M_\infty^{(i)}$ (over $GF(q^2)$) and intersecting O_i over $GF(q)$ in a nonsingular conic containing ∞, and the nonsingular elliptic quadrics $O \neq O_i$ for which $O \cap O_i$ over $GF(q^2)$ is $L_\infty^{(i)} \cup M_\infty^{(i)}$ (counted twice), with $1 \leq i \leq q-1$.

(b) The q^3 points of $PG(3,q) \setminus \pi_\infty$.

(c) The q^3 planes of $PG(3,q)$ not containing ∞.

(d) The $q-1$ sets $\mathcal{O}_i$, where $\mathcal{O}_i$ consists of the q^3 quadrics O of type (a) corresponding with O_i, $1 \leq i \leq q-1$.

(e) The plane π_∞.

(f) The point ∞.

Lines of $\mathcal{S}$

(i) Let (ω, γ) be a point-plane flag of $PG(3,q)$, with $\omega \notin \pi_\infty$ and $\infty \notin \gamma$. Then all quadrics O of type (a) which are tangent to γ at ω, together with ω and γ, form a line of type (i). Any two distinct quadrics of such a line have exactly two points (∞ and ω) in common. The total number of lines of type (i) is q^5.

(ii) Let O be a point of type (a) which corresponds to the quadric O_i, $1 \leq i \leq q-1$. If $O \cap \pi_\infty = O_i \cap \pi_\infty = L_\infty^{(i)} \cup M_\infty^{(i)}$ (over $GF(q^2)$), then all points O' of type (a) for which $O' \cap O$ over $GF(q^2)$ is $L_\infty^{(i)} \cup M_\infty^{(i)}$ counted twice, together with O and O_i, form a line of type (ii). There are $q^2(q-1)$ lines of type (ii).

(iii) A set of q parallel planes of $AG(3,q) = PG(3,q) \setminus \pi_\infty$, where the line at infinity does not contain ∞, together with the plane π_∞, is a line of type (iii).

(iv) Lines of type (iv) are the lines of $PG(3,q)$ not in π_∞ containing ∞.

(v) $\{\infty, \pi_\infty, \mathcal{O}_1, \mathcal{O}_2, \ldots, \mathcal{O}_{q-1}\}$ is the unique line of type (v).

Incidence of $\mathcal{S}$

Incidence is containment.

It is proved in [Th99] (with an additional hypothesis in case $q = 2^e$) that $\mathcal{S}$ is a GQ isomorphic to the point-line dual of the flock GQ constructed from the flock F. The approach of J. A. Thas is quite technical and uses results from algebraic geometry. His results led to a more elementary approach to the problem of giving a geometric construction for flock GQ in characteristic 2 initiated by Barwick, Brown and Penttila in [BBP06]. More recently, M. R. Brown has given a truly satisfactory treatment in the preprints [Br07] and [Br06]. His work culminates in a complete proof that a GQ with parameters (q^2, q) with Property (G) at a point must be isomorphic to a flock GQ.

9.3 A Result of N. L. Johnson

Let $q = p^e$ for an arbitrary prime p and suppose that $\mathcal{C} = \{A_t = \left(\begin{smallmatrix} xt & yt \\ 0 & zt \end{smallmatrix}\right) : t \in F\}$ is a q-clan. It was noted by N. L. Johnson [Jo87] that the associated translation planes are semifield planes if and only if coordinates can be chosen so that the map $t \mapsto A_t$ is an additive map. And it was shown in [Pa89] that the point-line dual of the associated flock generalized quadrangle is a translation generalized quadrangle (TGQ) if and only if the same condition holds. So it came as a bit of a surprise when N. L. Johnson [Jo87] proved the following theorem.

Theorem 9.3.1. *([Jo87]) If $q = 2^e$ and if $t \mapsto A_t$ is an additive map, then $\mathcal{C}$ is classical. Hence if $\mathcal{S}$ is a flock GQ whose point-line dual is a TGQ, then $\mathcal{S}$ is classical.*

Proof. The proof in [Jo87] is quite technical (if the computations quoted from earlier papers are included). The proof here is rather pretty and uses only the characterization by S. E. Payne [Pa71] of translation ovals. This says that if $t \mapsto t^\rho$ is an additive function for which $\mathcal{D}(\rho) = \{(1, t, t^\rho) : t \in F\} \cup \{(0,0,1)\}$ is an oval, then there is a nonzero scalar $c \in F$ and an automorphism β that generates the Galois group $\mathrm{Aut}(F)$ for which $t^\rho = ct^\beta$ for all $t \in F$. (This proof of Johnson's theorem was first shown to us by T. Penttila.)

So suppose that $\mathcal{C}$ as above is a q-clan for which the map $t \mapsto A_t$ is an additive map. For each $\alpha \in PG(1,q)$ we know that

$$\mathcal{O}_\alpha = \{(1, y_t, \alpha A_t \alpha^T) : t \in F\} \cup \{(0,1,0)\}$$

is an oval in $PG(2,q)$. By assumption $t \mapsto y_t = t^u$ is an additive (and nonsingular) map. Hence u^{-1} is also additive, implying that $t \mapsto A_{t^{u^{-1}}}$ is an additive map. Then we also have that

$$\mathcal{O}_\alpha = \{(1, t, \alpha A_{tu^{-1}}\alpha^T) : t \in F\} \cup \{(0,1,0)\}$$

is a translation oval for each $\alpha \in PG(1,q)$. By the theorem of [Pa71], for each $\alpha \in PG(1,q)$ the map $t \mapsto \alpha A_{tu^{-1}}\alpha^T$ must be of the form $t \mapsto at^\beta$ for β a generator of $\mathcal{A} = Gal(GF(q))$. For $\alpha = (1,0)$, $t \mapsto x_{tu^{-1}} = at^\sigma$ for $0 \neq a \in F$, σ a generator of $\mathcal{A}$. For $\alpha = (0,1)$, $t \mapsto z_{tu^{-1}} = ct^\tau$ for $0 \neq c \in F$, τ a generator of $\mathcal{A}$. Then for $\alpha = (1,1)$, $t \mapsto x_{tu^{-1}} + y_{tu^{-1}} + z_{tu^{-1}} = at^\sigma + t + ct^\tau = bt^\delta$ for $0 \neq b \in F$, δ a generator of $\mathcal{A}$. Since the elements of $\mathcal{A}$ are linearly independent over F, $\sigma = \tau = \delta = id$. Hence $A_{tu^{-1}} = \left(\begin{smallmatrix} at & t \\ 0 & ct \end{smallmatrix}\right)$, i.e., $A_t = t^u \left(\begin{smallmatrix} a & 1 \\ 0 & c \end{smallmatrix}\right)$, implying that $\mathcal{C}$ is classical. □

9.4 Translation (Hyper)Ovals and α-Flocks

Let $F = GF(q)$, $q = 2^e$. A *translation oval* $\mathcal{O}$ of $PG(2,q)$ is an oval for which there is a tangent line l such that the group of all elations with axis l and stabilising $\mathcal{O}$ acts regularly on $\mathcal{O} \setminus \{l\}$. In this case, l is called an *axis of* $\mathcal{O}$. A *translation hyperoval* is a hyperoval containing a translation oval. As is well known, the automorphisms of the field $GF(2^e)$ (namely, the maps $x \mapsto x^{2^i}$ for $0 \leq i \leq e-1$) form a cyclic group of order e generated by the Frobenius automorphism $x \mapsto x^2$. A generator of the cyclic group of automorphisms is said to be an *automorphism of maximal order* and it will be denoted by α. We remark that the hyperovals $\mathcal{D}(\alpha) = \{(1, t, t^\alpha) \colon t \in F\} \cup \{(0,1,0),(0,0,1)\}$, for α a generator of $Aut(F)$ are translation hyperovals of $PG(2,q)$, q even. In particular, conics are translation ovals, and every tangent line to a conic is an axis, and so regular hyperovals are translation hyperovals. The following theorem is due to Payne [Pa71].

Theorem 9.4.1. *Every translation hyperoval of* $PG(2,q)$ *is equivalent to* $\mathcal{D}(\alpha)$, *for some generator* α *of* $Aut(F)$.

In this section a generalization of the concept of a flock of a quadratic cone in $PG(3,q)$ is given. Let α be a generator of $Aut(F)$. Following [C98], an α*-cone* Σ_α in $PG(3,q)$ is defined to be a cone with base a translation oval in a plane π of $PG(3,q)$ and with vertex not belonging to π. In standard form, $\Sigma_\alpha = \{(x,y,z,w) \colon y^\alpha = xz^{\alpha-1}\}$ which is the α-cone with vertex $(0,0,0,1)$ and with base the translation oval $\{(t^\alpha, t, 1, 0) \colon t \in F\} \cup \{(1,0,0,0)\}$ in $\pi \colon w = 0$. An α*-flock* is a set of q planes π_t, $t \in F$, of $PG(3,q)$ not passing through the vertex of the α*-cone* which do not intersect each other at a point of Σ_α. (Alternatively, we could define an α*-flock* as a partition of the α*-cone* into q disjoint α*-conics*, where an α*-conic* is obtained as the intersection of the α-cone with each of the planes π_t). Following Thas [Th87], the planes π_t of an α-flock may be expressed by $a_t x + b_t y + c_t z + w = 0$, $t \in GF(q)$ (although Thas employed this formulation for a flock of a quadratic cone). An algebraic condition can be derived for q planes, not containing $(0,0,0,1)$, to be an α flock:

Theorem 9.4.2. *The set of planes $\{\pi_t \colon t \in GF(q)\}$ where π_t is given by the equation $a_t x + b_t y + c_t z + w = 0$, $t \in GF(q)$ is an α-flock if and only if*

$$trace\left(\frac{(a_t + a_s)^{\frac{1}{\alpha-1}}(c_t + c_s)}{(b_t + b_s)^{\frac{\alpha}{\alpha-1}}}\right) = 1 \quad \forall\, t, s \in GF(q),\ t \neq s. \tag{9.1}$$

We can also extend the definition of q-clan to the definition of α-clan in the following way. An α-clan is a set of q upper triangular matrices $A_t = \left(\begin{smallmatrix} at & bt \\ 0 & ct \end{smallmatrix}\right)$, $t \in F$ such that there exists a $k \in F$, with $trace(k) = 1$, so that

$$trace\left(k\frac{(a_t + a_s)^{\frac{1}{\alpha-1}}(c_t + c_s)}{(b_t + b_s)^{\frac{\alpha}{\alpha-1}}}\right) = 1 \quad \forall\, t, s \in GF(q),\ t \neq s. \tag{9.2}$$

Note that the inclusion of k in the definition of α-clan is a cosmetic change from the earlier definition of q-clans given in Chapter 1. Also note that the q-clans previously introduced are 2-clans according to this new definition. The following theorem is the analogue of Theorem 1.3.2.

Theorem 9.4.3. *$\{A_t \colon t \in F\}$ is an α-clan if and only if $\{\pi_t \colon t \in F\}$ is an α-flock.*

There are certain normalizations we can make in an α-clan which will make life easier in terms of computations. A convenient choice (without loss of generality) is to work with α-clans which are written in the following *standard form*:

$$A_t = \begin{pmatrix} f(t) & t^{1/\alpha} \\ 0 & g(t) \end{pmatrix}, \quad t \in F,$$

where f and g are permutation polynomials over F with $f(0) = g(0) = 0$. Note that the condition that the A_t's in this form give an α-clan is that there exists a $k \in F$, with $trace(k) = 1$, such that

$$trace\left[k\left(\frac{(f(t) + f(s))}{t + s}\right)^{\frac{1}{\alpha-1}}(g(t) + g(s))\right] = 1 \quad \forall\, t, s \in GF(q),\ t \neq s. \tag{9.3}$$

An interesting and important property of α-flocks is the following connection with hyperovals of $PG(2, q)$.

Theorem 9.4.4. *If $\{A_t = \left(\begin{smallmatrix} f(t) & t^{1/\alpha} \\ 0 & g(t) \end{smallmatrix}\right) : t \in F\}$ is an α-clan, then $f(t)$ is an o-polynomial.*

Among the known hyperovals only the O'Keefe–Penttila hyperoval of $PG(2, 32)$ (see [OKP92]) does not arise from an α-flock for any α (see [BOPPR2]). The concept of α-flock turns out to be very useful to prove that the Cherowitzo hyperovals form an infinite family of hyperovals. The Cherowitzo hyperovals are the following:

$$\mathcal{O}^+ = \{(1, t, t^\sigma + t^{\sigma+2} + t^{3\sigma+4}) \colon t \in F\} \cup \{(0, 1, 0), (0, 0, 1)\}.$$

What follows is an outline of how α flocks are used to show that result. The concept of 'flippability' is needed. An α-clan $\begin{pmatrix} f(t) & t^{1/\alpha} \\ 0 & g(t) \end{pmatrix}$ is called $\rho(t)$-*flippable* for some permutation $\rho(t)$ of F with $\rho(0) = 0$ provided that $\begin{pmatrix} \frac{f(t)}{\rho(t)} & \frac{t^{1/\alpha}}{\rho(t)} \\ 0 & \frac{g(t)}{\rho(t)} \end{pmatrix}$ is also an α-clan with respect to the same *trace* 1 element k, called the $\rho(t)$-*flip* of the original. Note that the concept of flippability is a generalization of the operation on 2-clans called 'flip' we have already encountered before.

Lemma 9.4.5. *If* $q = 2^e = 2^{2h+1}$, *e odd then* $\left\{\begin{pmatrix} 1+(1+t)^{\sigma+2} & t^{\sigma} \\ 0 & 1+(1+t)^{3\sigma+4} \end{pmatrix} : t \in F\right\}$ *is a t-flippable* $1/\sigma$*-clan, where* σ *is the automorphism* $x \mapsto x^{2^h}$.

Using now Theorem 9.4.4 and Lemma 9.4.5, we get

Theorem 9.4.6. $f(x) = x^{\sigma} + x^{\sigma+2} + x^{3\sigma+4}$ *is an o-polynomial over* $GF(2^e)$ *for all odd e.*

9.5 Monomial hyperovals

A *monomial hyperoval* of $PG(2,q)$ is one equivalent to $\mathcal{D}(x^k) = \{(1,x,x^k) : t \in GF(q)\} \cup \{(0,1,0),(0,0,1)\}$, for some integer k.

Recall that $\mathcal{D}(x^k)$, $\mathcal{D}(x^{1-k})$, $\mathcal{D}(x^{1/k})$, $\mathcal{D}(x^{1-1/k})$, $\mathcal{D}(x^{k/(k-1)})$, $\mathcal{D}(x^{1/(1-k)})$, with $1-k$, $1/k$, $1-1/k$, $k/(k-1)$, $1/(1-k)$ calculated *mod* $(q-1)$ are projectively equivalent. The known examples are the *translation hyperoval* $\mathcal{D}(x^{2^i})$ in $PG(2,2^h)$, $(i,h) = 1$, the *Segre* hyperoval $\mathcal{D}(x^6)$ in $PG(2,2^h)$, h odd, and the two *Glynn* hyperovals $\mathcal{D}(x^{\sigma+\gamma})$ and $\mathcal{D}(x^{3\sigma+4})$ in $PG(2,q)$, $q = 2^h$, h odd, $\sigma^2 \equiv 2 \ mod\,(q-1)$, $\gamma^2 \equiv \sigma \ mod\,(q-1)$, with σ, γ automorphisms of F. By using a computer, it has been checked up to $q = 2^{30}$ that no other monomial hyperovals exist [Gl85]. This has led to the conjecture that all monomial hyperovals have been determined.

A Result of W. E. Cherowitzo and L. Storme

The following theorem by Cherowitzo and Storme (see [CS98]) supports that conjecture and gives significant progress on the classification of monomial hyperovals.

Theorem 9.5.1. $\mathcal{D}(x^k)$ *with* $k = 2^i + 2^j$, *with* $i \neq j$ *is a hyperoval in* $PG(2,2^h)$ *if and only if* $h = 2^e - 1$ *is odd, and one of the following holds:*

1. $k = 2 + 2^2 = 6$, *Segre–Bartocci hyperoval;*
2. $k = \sigma + 2$ *with* $\sigma^2 \equiv 2\,(mod\, q-1)$, $\sigma \in Aut(F)$, *translation hyperoval;*
3. $k = \gamma + \gamma^2 = \gamma + \sigma\,(mod\, q-1)$ *with* $\gamma^4 \equiv 2\,(mod\, q-1)$, $\gamma \in Aut(F)$, *Glynn hyperoval;*
4. $k = q/4 + q/2 = 3/4$, *translation hyperoval.*

A Result of Penttila and Storme

We have already seen that a flock of the quadratic cone is equivalent to a herd of ovals in $PG(2,q)$, q even and to a flock-generalized quadrangle of order (q^2,q). O'Keefe and Penttila [OP] studied the problem of herds containing a translation oval. This is equivalent to studying the flock GQs which contain a subGQ $T_2(\mathcal{O})$ with $\mathcal{O}$ a translation oval. The following is the equivalent description of their result.

Theorem 9.5.2. (1) *If at least one oval in a herd is a translation oval, then this herd is either classical or FTWKB.*

(2) *If at least one subquadrangle $T_2(\mathcal{O})$ in a flock GQ is defined by a translation oval, then this flock GQ is the classical flock GQ $H(3,q^2)$ or the FTWKB-flock GQ.*

Penttila and Storme [PS98] studied herds containing a monomial oval which is not a translation oval. To do that they came to a complete classification of the monomial flocks as the following theorem states.

Theorem 9.5.3. *A monomial flock of the quadratic cone in characteristic 2 is linear, Fisher–Thas–Walker or Payne. A non-translation monomial oval in a herd is Segre–Bartocci.*

9.6 Conclusion and open Problems

We are aware that the material in this chapter does not cover all the vast subject on ovoids, spreads and flocks for q even. Our intention was to sketch some of the basic results and give references for the interested reader. We also think that in particular two of the previous sections deserve special attention: the geometric construction by Thas and the α-flocks of Cherowitzo. We believe that they provide interesting first steps which would lead to further developments of the theory.

Here we list some of the open problems we believe are worthy of further investigation:

1. Determine whether or not there are any other m that give cyclic GQ or even just give ovals (Sec. 4.6).
2. In all known cases each automorphism of one oval of a herd can be lifted to an automorphism of $G^{\otimes}$ and then to a collineation of the GQ. Give a unified proof of this.
3. Determine when an automorphism of a subquadrangle $T_2(\mathcal{O})$ embedded in a flock GQ $\mathcal{S}$ lifts to an automorphism of $\mathcal{S}$ (Sec. 1.8).

Bibliography

[BLP94] L. Bader, G. Lunardon and S. E. Payne, On q-clan geometry, $q = 2^e$, *Bull. Belgian Math. Soc., Simon Stevin* **1** (1994), 301–328.

[BLT90] L. Bader, G. Lunardon and J. A. Thas, Derivation of flocks of quadratic cones, *Forum Math.* **2** (1990), 163–174.

[BaBo71] A. Barlotti and R. C. Bose, Linear representation of a class of projective planes in four dimensional projective spaces, *Ann. Mat. Pura Appl.* **88** (1971), 9–31.

[BBP06] S. G. Barwick, M. R. Brown and T. Penttila, Flock generalized quadrangles and tetradic sets of elliptic quadrics of $PG(3, q)$. *J. Combin. Theory* Ser. A **113** (2006), no. 2, 273–290.

[Br00] M. R. Brown, Ovoids of $PG(3, q)$, q even, with a conic section, *J. London Math. Soc.* (2) **62** (2000), 569–582.

[Br07] M. R. Brown, Projective ovoids and generalized quadrangles, *Advances in Geometry* **7** (2007), 65–81.

[Br06] M. R. Brown, Laguerre geometries and some connections to generalized quadrangles, preprint.

[BOPPR1] M. R. Brown, C. M. O'Keefe, S. E. Payne, T. Penttila and G. F. Royle, Spreads of $T_2(\mathcal{O})$, α-flocks and ovals, *Designs, Codes and Cryptography* **31** (2004), 251–282.

[BOPPR2] M. R. Brown, C. M. O'Keefe, S. E. Payne, T. Penttila and G. F. Royle, The classification of spreads of $T_2(\mathcal{O})$ and α-flocks over small fields, in preparation.

[Bu95] F. Buekenhout, editor, *HANDBOOK OF INCIDENCE GEOMETRY: Buildings and Foundations*, North Holland, 1995, xi + 1420 pp.

[C98] W. E. Cherowitzo, α-flocks and hyperovals, *Geom. Ded.* **72** (1998), 221–146.

[C02] W. E. Cherowitzo, Flocks of cones: herds and herds spaces, preprint, 2002.

[CP03] W. E. Cherowitzo and S. E. Payne, The cyclic q-clans with $q = 2^e$, *Advances in Geometry*, Special Issue (2003), 158–185.

[CS98] W. E. Cherowitzo and L. Storme, α-flocks with oval herds and monomial hyperovals, *Finite fields and their Applications* **4** (1998), 185–199.

[COP03] W. E. Cherowitzo, C. M. O'Keefe, and T. Penttila, A unified construction of finite geometries related to q-clans in characteristic two, *Advances in Geometry* **3** (2003), 1–21.

[CPPR96] W. Cherowitzo, T. Penttila, I. Pinneri, G. Royle, Flocks and ovals, *Geom. Ded.* **60** (1996), 17–37.

[DCGT88] F. De Clerck, H. Gevaert, and J. A. Thas, Flocks of a quadratic cone in $PG(3,q)$, $q \leq 8$, *Geom. Dedicata* **26** (1988), 215–230.

[DCH93] F. De Clerck and C. Herssens, Flocks of the quadratic cone in $PG(3,q)$, for q small, *Reports of the CAGe project* **8** (1993), 1–75.

[De68] P. Dembowski, *Finite Geometries*, Springer-Verlag, 1968.

[Di58] L. E. Dickson, *Linear Groups*, Dover Publications, Inc., New York, 1958.

[FT79] J. C. Fisher and J. A. Thas, Flocks in $PG(3,q)$, *Math. Zeit.* **169** (1979), 1–11.

[GJ88] H. Gevaert and N. L. Johnson, Flocks of quadratic cones, generalized quadrangles and translation planes, *Geom. Dedicata*, **27** (1988), 301–317.

[GJT88] H. Gevaert, N. L. Johnson and J. A. Thas, Spreads covered by reguli, *Simon Stevin* **62** (1988), 51–62.

[Gl85] D. G. Glynn, A condition for the existence of ovals in $PG(2,q)$, q even, *Geom. Dedicata* **32** (1985), 247–252.

[Gl98] D. G. Glynn, Plane representation of ovoids, *Bull. Belg. Math. Soc. Simon Stevin* **5** (1998), 275–286.

[Ha75] M. Hall, Jr., Ovals in the desarguesian plane of order 16, *Annali Mat. Pura Appl.* **102** (1975), 159–176.

[Hi79] J. W. P. Hirschfeld, *Projective Geometries over Finite Fields*, Oxford University Press, Oxford (1979).

[Hi98] J. W. P. Hirschfeld, *Projective Geometries over Finite Fields*, 2nd Ed., Oxford University Press, Oxford (1998).

[Hu67] B. Huppert, Endliche Gruppen I, Springer-Verlag, 1967.

[Jo87] N. L. Johnson, Semifield flocks of Quadratic cones, *Simon Stevin, A Quarterly of Pure and Applied Mathematics* **1** (1987), Number 3–4.

[Jo89] N. L. Johnson, The derivation of dual translation planes, *J. Geom* **36** (1989), 63–90.

[Jo92] N. L. Johnson, Derivation of partial flocks of quadratic cones, *Rend. Mat.* **12** (1992), 817–848.

[JL94] N. L. Johnson and G. Lunardon, On the Bose-Barlotti Δ-planes, *Geom. Dedicata* **49** (1994), 173–182.

[JLW91] N. L. Johnson, G. Lunardon and F. W. Wilke, Semifield skeletons of conical flocks, *J. Geom.* **40** (1991), 105–112.

[JP97] N. L. Johnson and S. E. Payne, Flocks of Laguerre Planes and Associated Geometries. Mostly Finite Geometries (Ed. N. L. Johnson), Marcel Dekker, 1997, pp. 51–122.

[Ka79] W. M. Kantor, Classical groups from a non-classical viewpoint, Mathematical Institute (Oxford Lectures 1978), Oxford, 1979.

[Ka80] W. M. Kantor, Generalized quadrangles associated with $G_2(q)$, *Jour. Comb. Theory (A)* **29** (1980), 212–219.

[Ka86] W. M. Kantor, Some generalized quadrangles with parameters (q^2, q), *Math. Zeit.* **192** (1986), 45–50.

[Kn92] N. Knarr, A geometric construction of generalized quadrangles from polar spaces of rank three, *Results Math.* **21** (1992), no. 3–4, 332–344.

[Ko78] G. Korchmáros, Gruppi di collineazioni transitivi sui punti di una ovale [$(q+2)$-arco] di $S_{2,q}$, q pari, *Atti Sem. Mat. Fis. Univ. Modena* **27** (1978), 89–105.

[LS58] L. Lunelli and M. Sce, k-archi completi nei piani proiettivi desarguesiani di rango 8 e 16, 1958. Centro di Calcoli Numerici, Politecnico di Milano.

[OKP91] C. M. O'Keefe and T. Penttila, Hyperovals in $PG(2,16)$, *European Jour. Combin.* **12** (1991), 51–59.

[OKP92] C. M. O'Keefe and T. Penttila, A new hyperoval in $PG(2,32)$, *Journ. Geom.* **44** (1992), 117–139.

[OP] C. M. O'Keefe and T. Penttila, Characterization of flock quadrangles, *Geom. Dedicata* **82** (2000), 171–191.

[OP02] C. M. O'Keefe and T. Penttila, Automorphism groups of generalized quadrangles via an unusual action of $P\Gamma L(2, 2^h)$, *European J. Comb.* **23** (2002), no. 2, 213–232.

[OPP91] C. M. O'Keefe, T. Penttila and C. E. Praeger, Stabilizers of hyperovals in $PG(2,32)$, *Advances in finite geometries and designs* (Chelwood Gate, 1990), 337–351, Oxford Sci. Publ., Oxford Univ. Press, New York, 1991.

[OKT96] C. M. O'Keefe and J. A. Thas, Collineations of Subiaco and Cherowitzo hyperovals, *Bull. Belg. Math. Soc. Simon Stevin* **3** (1996), 177–192.

[Pa71] S. E. Payne, A complete determination of translation ovoids in finite Desarguesian planes, *Atti Accad. Naz. Lincei Rend. Cl. Sci. Fis. Mat. Natur.* (8) **51** (1971), 328–331.

[Pa80] S. E. Payne, Generalized quadrangles as group coset geometries, *Congr. Numer.* **29** (1980), 717–734.

[Pa85] S. E. Payne, A new infinite family of generalized quadrangles, *Congr. Numer.* **49** (1985), 115–128.

[Pa86] S. E. Payne, Hyperovals yield many GQ, *Simon Stevin* **60** (1986), 211–225.

[Pa89] S. E. Payne, An essay on skew translation generalized quadrangles, *Geom. Dedicata* **32** (1989), 93–118.

[Pa92] S. E. Payne, Collineations of the generalized quadrangles associated with q-clans, *Ann. Discr. Math.* **52** (1992), 449–461.

[Pa94] S. E. Payne, Collineations of the Subiaco generalized quadrangles, *Bull. Belgian Math. Soc., Simon Stevin* **1** (1994), 427–438.

[Pa95] S. E. Payne, A tensor product action on q-clan generalized quadrangles with $q = 2^e$, *Lin. Alg. and its Appl.* **226–228** (1995), 115–137.

[Pa96] S. E. Payne, The fundamental theorem of q-clan geometry, *Designs, Codes and Cryptography* **8** (1996), 181–202.

[Pa98] S. E. Payne, *The Subiaco Notebook: An Introduction to q-Clan Geometry,* $q = 2^e$, available at `http://www-math.cudenver.edu/~spayne/`.

[Pa02a] S. E. Payne, Four Lectures in Napoli: An Introduction to q-Clan Geometry, $q = 2^e$, Winter, 2002, 63 pages.

[Pa02b] S. E. Payne, The Cyclic q-Clans with $q = 2^e$, May 2002, unpublished preprint, 26 pages.

[PC78] S. E. Payne and J. E. Conklin, An unusual generalized quadrangle of order sixteen, *Jour. of Comb. Theory* (A) **24** (1978), 50–74.

[PM82] S. E. Payne and C. C. Maneri, A family of skew-translation generalized quadrangles of even order, *Congress. Numer.* **36** (1982), 127–135.

[PPP95] S. E. Payne, T. Penttila, I. Pinneri, Isomorphisms between Subiaco q-clan geometries, *Bull. Belgian Math. Soc.*, **2** (1995), 197–222.

[PPR97] S. E. Payne, T. Penttila, G. F. Royle, Building a cyclic q-clan, *Mostly Finite Geometries*, (ed. N. L. Johnson), Marcel Dekker, 1997, 365–378.

[PR90] S. E. Payne and L. A. Rogers, Local group actions on generalized quadrangles, *Simon Stevin* **64** (1990), 249–284.

[PT84] S. E. Payne and J. A. Thas, *Finite generalized quadrangles*, Pitman, 1984.

[PT91] S. E. Payne and J. A. Thas, Conical flocks, partial flocks, derivation and generalized quadrangles, *Geom. Dedicata* **38** (1991), 229–243.

[PKT02] S. E. Payne and K. Thas, Notes on elation generalized quadrangles, *European Journal of Combinatorics* **24** (2003), 969–981.

[PT05] S. E. Payne and J. A. Thas, The stabilizer of the Adelaide oval, *Discrete Mathematics* **294** (2005), 161–173.

[Pe99] T. Penttila, A plane representation of ovoids, in *Util. Math.* **56** (1999), 245–250.

[Pe02] T. Penttila, Configurations of ovals, in *Combinatorics 2002: Topics in Combinatorics: geometry, Graph theory and Designs*, Maratea (Potenza), Italy, 2–8 June 2002, pp. 220–241.

[PP94] T. Penttila and I. Pinneri, Irregular hyperovals in $PG(2, 64)$, *Jour. Geom.* **51** (1994), 89–100.

[PR95] T. Penttila and G. Royle, On hyperovals in small projective planes, *Jour. Geom.* **54** (1995), 91–104.

[PS98] T. Penttila and L. Storme, Monomial flocks and herds containing a monomial oval, *J. Combin. Theory*, Ser. A **83** (1998), 21–41.

[Se57] B. Segre, Sui k-archi nei piani finiti di caratteristica due, *Rev. Mat. Pures Appl.* **2** (1957), 289–300.

[Se62] B. Segre, Ovali e curve σ nei piani di Galois di caratteristica due., *Atti Accad. Naz. Lincei Rend. Cl. Sci. Fis. Mat. Nat.*, **8** (1962),785–790.

[Se68] A. Seidenberg, *Elements of the Theory of Algebraic Curves*, Addison-Wesley, Reading, Mass., 1968, 216 pp.

[ST95] L. Storme and J. A. Thas, k-Arcs and partial flocks, *Linear Algebra Appl.* **226** (1995), 33–45.

[Th87] J. A. Thas, Generalized quadrangles and flocks of cones, *European J. Combin.*, **8** (1987), 441–452.

[Th93] J. A. Thas, A characterization of the Fisher-Thas-Walker flocks, *Simon Stevin* **3–4** (1993), 219–226.

[Th98] J. A. Thas, 3-regularity in generalized quadrangles: a survey, recent results and the solution of a longstanding conjecture, *Rend. Circ. Math. Palermo* (2) *Suppl.*, 53 (1998), 199–218. Combinatorics '98 (Mondello).

[Th99] J. A. Thas, Generalized quadrangles of order (s, s^2), III, *J. Combin. Theory Ser. A* **87** (1999), 247–272.

[Th01a] J. A. Thas, Geometrical constructions of flock generalized quadrangles, *J. Combin. Theory Ser. A* **94** (2001), 51–62.

[Th01b] J. A. Thas, A result on spreads of the generalized quadrangle $T_2(\mathcal{O})$, with $\mathcal{O}$ an oval arising from a flock, and applications, *European J. Combin.* **22** (2001), 879–886.

[THDC93] J. A. Thas, C. Herssens and F. De Clerck, and Flocks and partial flocks of the quadratic cone in $PG(3, q)$, in: *Finite Geometry and Combinatorics*, Vol. 77, Cambridge University Press, Cambridge, 1993, 379–393.

[TP94] J. A. Thas and S. E. Payne, Spreads and ovals in finite generalized quadrangles, *Geom. Dedicata* **52** (1994), 227–253.

[TPG88] J. A. Thas, S. E. Payne and H. Gevaert, A family of ovals with few collineations, *Europ. J. Combin.* **9** (1988), 353–362.

[TTVM06] J. A. Thas, K. Thas and H. Van Maldeghem *Translation Generalized Quadrangles*, Series in Pure Mathematics, World Scientific, 2006.

[Ti59] J. Tits, Sur la trialité et certains groupes qui s'en déduisent, Inst. Hautes Etudes Sci. Publ. Math. **2** (1959), 14–60.

[Wa76] M. Walker, A class of translation planes, *Geom. Dedicata* **5** (1976), 135–146.

Index